SpringerBriefs in Complexity

SpringerBriefs in Complexity are a series of slim high-quality publications encompassing the entire spectrum of complex systems science and technology. Featuring compact volumes of 50 to 125 pages (approximately 20,000–45,000 words), Briefs are shorter than a conventional book but longer than a journal article. Thus Briefs serve as timely, concise tools for students, researchers, and professionals.

Typical texts for publication might include:

- A snapshot review of the current state of a hot or emerging field
- A concise introduction to core concepts that students must understand in order to make independent contributions
- An extended research report giving more details and discussion than is possible in a conventional journal article,
- A manual describing underlying principles and best practices for an experimental or computational technique
- An essay exploring new ideas broader topics such as science and society

Briefs allow authors to present their ideas and readers to absorb them with minimal time investment. Briefs are published as part of Springer's eBook collection, with millions of users worldwide. In addition, Briefs are available, just like books, for individual print and electronic purchase. Briefs are characterized by fast, global electronic dissemination, straightforward publishing agreements, easy-to-use manuscript preparation and formatting guidelines, and expedited production schedules. We aim for publication 8–12 weeks after acceptance.

SpringerBriefs in Complexity are an integral part of the Springer Complexity publishing program. Proposals should be sent to the responsible Springer editors or to a member of the Springer Complexity editorial and program advisory board (springer.com/complexity).

More information about this series at http://www.springer.com/series/8907

Shamik Gupta · Alessandro Campa
Stefano Ruffo

Statistical Physics
of Synchronization

 Springer

Shamik Gupta
Department of Physics
Ramakrishna Mission Vivekananda
    University
Howrah, India

Alessandro Campa
National Center for Radiation Protection
    and Computational Physics
Istituto Superiore di Sanità
Roma, Italy

Stefano Ruffo
SISSA
Trieste, Italy

ISSN 2191-5326   ISSN 2191-5334   (electronic)
SpringerBriefs in Complexity
ISBN 978-3-319-96663-2   ISBN 978-3-319-96664-9   (eBook)
https://doi.org/10.1007/978-3-319-96664-9

Library of Congress Control Number: 2018949061

This Springer imprint is published by the registered company Springer Nature Switzerland AG
The registered company address is: Gewerbestrasse 11, 6330 Cham, Switzerland

To my beloved parents,
Late Subir Kumar Gupta and Geetanjali
Gupta

Shamik Gupta

To the memory of my mother, Maria

Alessandro Campa

To Antonella

Stefano Ruffo

Foreword

In 1965, a bright undergraduate at Cornell University named Arthur Winfree undertook an experimental study for his senior thesis. He wired 71 neon tube oscillators together into a contraption that he called the firefly machine. At a time when the theory of nonlinear oscillations was largely confined to two and three oscillators, Winfree was venturing out to study dozens of them. To allow the oscillators to feel one another's influence, he connected them all to a common terminal through a small capacitor, so that each oscillator interacted equally with all the others, and he grounded that terminal through a larger variable capacitor. This setup enabled him to easily adjust the coupling strength between the oscillators.

He found that in the absence of coupling, the neon tubes blinked on and off in an uncoordinated, incoherent fashion. That was to be expected; their natural periods varied by about 10%. As Winfree slowly dialed up the coupling, the oscillators remained incoherent until it reached a critical coupling strength. Above that threshold, all the neon tubes began discharging in unison, much like the famous congregations of synchronously flashing fireflies of Thailand, Malaysia, and other parts of Southeast Asia. Winfree had discovered the sudden onset of synchronization in a population of nonlinear oscillators.

Winfree's inspiration had always been biology—not just fireflies with their rapid flash rhythms, but the much slower rhythms in sleep and wake and body temperature of mammals, the nearly 24-hour rhythms known as circadian rhythms. The early 1960s were the heyday of research into circadian rhythms. With his firefly machine, Winfree opened up a theoretical avenue for studying such rhythms.

At the time, Winfree was a college student majoring in engineering physics. His training in solid-state theory led him to approach the question of biological synchronization from a perspective that only a physicist would have. He realized that an infinite-range approximation, in which each oscillator interacted equally with all the others, offered the best hope of making progress on this daunting, nonlinear, nonequilibrium, many-body problem. That was why he coupled all the oscillators through a common capacitor. He was doing the electronic counterpart of mean-field theory.

Next, Winfree abstracted his firefly machine into a set of differential equations that he could simulate on the university's mainframe computer. At the time, computer simulations were a rarity in science. One had to go to a computing center and feed punch cards into a room-sized behemoth. To simplify the differential equations, Winfree assumed that his model oscillators were weakly coupled, compared to their attraction to their limit cycles in state space. He realized intuitively that under that assumption, each oscillator could be represented by its phase alone as it moved along its limit cycle; amplitude variations could be neglected. In a now-celebrated paper published in *Journal of Theoretical Biology* in 1967, Winfree showed that his mathematical model could do what his firefly machine had done: it could spontaneously synchronize. As the coupling strength between the oscillators was increased, or as the variance of oscillators' natural frequencies was decreased, the oscillators abruptly switched from an incoherent, desynchronized state to an ordered state in which a macroscopic fraction of the system was locked in frequency. In this 1967 paper, he explicitly noted a remarkable connection to thermodynamic phase transitions. He wrote:

> Disguised in the literature of solid-state physics under an interchange of spatial for temporal coordinates, the phenomenon of ferroelectric crystallization is strikingly analogous: the oscillators are replaced by a population of electric dipoles at crystal lattice points; the orientation of their phase vectors [...] becomes the angular orientation of dipoles under a communally-generated electric field, to which they contribute [...] according to orientation; the spread of synchronized phases [...] due to the spread of natural frequencies [...] becomes the distribution of dipole angles due to thermal buffeting; and the threshold [of synchronization] is mirrored in the Curie temperature for ferroelectric transition.

About a decade later, the Japanese statistical physicist Yoshiko Kuramoto reformulated Winfree's work and recast it as a beautifully elegant system of differential equations, now known as the Kuramoto model. Using an ingenious self-consistency argument, and retaining Winfree's assumptions of a mean-field model of phase-only oscillators, but using the more tractable form of coupling between the oscillators, Kuramoto was able to find his synchronization transition analytically and to calculate the extent of order above the synchronization threshold.

In the half a century since Winfree's landmark work, the study of collective synchronization has mainly been approached through nonlinear dynamics and computer simulation. The connection to statistical physics, though always present, has tended to play a subordinate role. The present monograph rectifies this situation. Shamik Gupta, Alessandro Campa, and Stefano Ruffo do a wonderful job of summarizing earlier work on the Kuramoto model and enlarging it to embrace the insights of statistical physics, using concepts like H-theorems, Fokker–Planck equations, and the breakdown of detailed balance. The problems they tackle are difficult and fascinating, both from the standpoint of nonlinear dynamics and from that of statistical physics, because of their nonequilibrium and many-body character. Furthermore, the authors explore the effects of inertia, always an important physical consideration, but one that has been given relatively little attention in the nonlinear dynamics literature. This is a very valuable addition to the literature of dynamical systems and nonequilibrium statistical physics. I hope you'll enjoy reading it as much as I did.

Ithaca, New York, USA

Prof. Steven Strogatz
Department of Mathematics
Cornell University

Preface

A remarkable phenomenon common in nature is that of spontaneous synchronization, whereby a large population of coupled oscillating units of diverse frequencies spontaneously evolve to operate in unison. Such a cooperative effect commonly occurs in physical and biological systems over length and time scales of several orders of magnitude. Examples are flashing of fireflies, animal flocking behavior, audience clapping in concert halls, pedestrians on footbridges, and a variety of experiments involving electrochemical and electronic oscillators, metronomes, Josephson junctions, and laser arrays. Besides its necessity in firing of cardiac cells that keeps the heart beating and life going, synchrony is desired in man-made systems, e.g., in parallel computing, whereby computer processors must coordinate to finish a task on time, and in electrical power grids, whereby generators must run in synchrony to be locked in frequency to that of the grid. Synchrony could also be hazardous, e.g., in neurons, leading to impaired brain functions in, e.g., epilepsy. Collective synchrony among oscillators has attracted immensely the attention of physicists and applied mathematicians, and finds applications in many fields, from quantum electronics to electrochemistry, from bridge engineering to social science, and others.

Synchronizing systems may be viewed from two contrasting perspectives, namely, that of dynamical systems theory and statistical physics. To summarize in one sentence the characterizing aspects of the two perspectives, one could say that in the former, spontaneous synchronization occurs as a bifurcation in the dynamical behavior of the system as a function of the strength of interaction between the oscillating units, while in the latter, it represents a phase transition between different forms of statistical distribution of the dynamical variables of the system constituents. Until now, the approach based on dynamical systems theory has received much more attention. This could be partly due to the fact that synchronizing systems have mostly been investigated using models not belonging to any class of Hamiltonian systems, the latter constituting the prominent subject of study in mainstream statistical physics. The use of mainly models with first-order dynamics (only very recently are models with second-order dynamics being studied) has been one other reason for the abundance of studies employing tools of dynamical systems theory.

Viewed from the perspective of statistical physics, the following characteristics of synchronizing systems may be noted. Presence of long-range interactions in synchronizing systems allows the use of mean-field models, which may be seen as a major simplifying feature for extensive analytical treatments. The mean-field analysis becomes exact in the limit of a very large number of units (in particular, in the thermodynamic limit, which is naturally achieved in synchronizing systems) for systems where the interaction is the same between every pair of constituents. The latter feature is not always prevalent in real systems, as there are cases where the interaction, although long-ranged, decays slowly with the distance between the constituents; nevertheless, also in this case, the mean-field analysis is a very useful first approximation, and corrections can in principle be evaluated systematically. Another essential feature of synchronizing systems is the presence of diverse natural frequencies. In the language of statistical physics, diverse frequencies may be interpreted as quenched disordered random variables; the randomness implies the necessity to average observable quantities over the distribution of natural frequencies. Probably, the most notable feature of synchronizing systems is the fact that the stationary states to which the dynamics settles to after a transient are not equilibrium ones (in technical terms, such states do not satisfy detailed balance). Thus, synchrony is necessarily a nonequilibrium phenomenon, which therefore cannot be described by equilibrium statistical mechanics. There is as yet no theory akin to the latter that can treat and make predictions on general terms for nonequilibrium systems, thus necessitating the study of representative model systems so as to gain valuable insights into the physics of synchronizing systems. Summarizing, synchronizing systems involve the study of statistical physics of long-range systems with quenched random variables settling into nonequilibrium steady states. This brief monograph aims to present from this perspective a study of synchronizing systems.

Extensive studies of synchronizing systems over the years have led to the introduction of novel theoretical concepts in nonlinear science such as the chimera states. Chimeras are broken-symmetry states occurring in identical, symmetrically coupled oscillator ensembles in which synchronized and desynchronized subpopulations coexist. These states have been observed in a variety of experimental situations involving, e.g., chemical and mechanical oscillators. Dynamical phenomena such as chimeras have been studied analytically using the approach of dynamical systems theory. Our focus in this monograph is on statistical physics approach to synchronization, and interpreting chimeras, etc., within this approach is still largely an open issue. Hence, we will not dwell on such dynamical phenomena, interesting in their own right, in this brief monograph.

It is a great pleasure to warmly thank a number of colleagues for fruitful and enjoyable discussion and collaboration on topics discussed in this monograph: Eduardo G. Altmann, Julien Barré, Freddy Bouchet, Lapo Casetti, Pierfrancesco Di Cintio, Thierry Dauxois, Stefano Gherardini, Maxim Komarov, Haggai Landa, David Métivier, Giovanna Morigi, David Mukamel, Hyunggyu Park, Arkady Pikovsky, Antonio Politi, Max G. Potters, Alessandro Torcini, Hugo Touchette, and Lucas Wetzel. SG is thankful to his parent organization, the Ramakrishna Mission Vivekananda University, for providing a conducive ambiance for writing this book.

Finally, we acknowledge the kind support of Springer-Verlag, Berlin. We are especially thankful to Aldo Rampioni and Kirsten Theunissen for their encouragement, assistance, and patience.

Book Homepage: Comments on the book may be sent to these email addresses: alessandro.campa@iss.it; shamik.gupta@rkmvu.ac.in; ruffo@sissa.it. Misprints and errors in the book will be posted on the webpage https://sites.google.com/site/shamikguptaphysics/syncbook.

Kolkata, India   Shamik Gupta
Roma, Italy   Alessandro Campa
Trieste, Italy   Stefano Ruffo
May 2018

Contents

Chapter 1
Synchronizing Systems

Abstract In the first section, we give a concise introduction to synchronizing systems, followed by a qualitative discussion in the next section of their representation in terms of interacting limit-cycle oscillators. In sections three and four, we discuss how each oscillating unit, either in isolation or in interaction with other units, may be effectively described with a phase variable having a first-order dynamics in time, and then deriving the form of interaction in terms of differences of phases between the oscillators. We also introduce the celebrated Kuramoto model, whose study constitutes the bulk of this monograph. In sections five and six, we consider in turn an interpretation of general synchronizing systems as statistical mechanics systems, and a discussion of tools and advantages of such an interpretation. In section seven, we then briefly derive analytical results on emergence of synchronization in the Kuramoto model, using the methods of Kuramoto, and Ott and Antonsen. In the final section, we address the behavior of the dynamics of oscillators with second-order dynamics in time, which is achieved when the oscillators have finite inertia; we provide a comparison of the behavior of isolated oscillators with and without inertia to demonstrate how the former may introduce significant effects on the dynamics.

Keywords Synchronization · Limit-cycle oscillators · Phase description
Oscillators in interaction · Kuramoto model
Statistical mechanics interpretation · Kuramoto and Ott-Antonsen solution
Effects of inertia

1.1 Introduction

Spontaneous collective synchronization is a remarkable phenomenon commonly observed in nature, whereby a sufficiently large population of single units that have themselves an oscillatory behavior tend to adjust their rhythm to oscillate at a common frequency. Indeed, the term *synchronization* arises from a combination of two Greek words, namely, *syn*, meaning same/common, and *chronos*, meaning time. A commonplace example of synchrony is observed in a group of dancers executing the same series of movements. The choreography could require that all dancers do the

© The Author(s) 2018
S. Gupta et al., *Statistical Physics of Synchronization*,
SpringerBriefs in Complexity, https://doi.org/10.1007/978-3-319-96664-9_1

same movement at the same time, or, that the second moves with a short time lag with respect to the first, the third with the same time lag with respect to the second, and so on. In either case, we would have no hesitation in concluding that the dancers are synchronized. The latter is true regardless of the specific movement that each dancer performs. In this book, we are interested in a particular kind of dynamics of single units, namely, a periodic dynamics. In the aforementioned example of dancers, such a dynamics would be represented by a choreography in which, e.g., each dancer repeats many times the same movement.

In a dynamical system undergoing a periodic motion, each dynamical variable attains the same value at intervals of time T called the period of the periodic motion. A paradigmatic example is that of the undamped harmonic oscillator, in which the position and the velocity vary in time with a period T characteristic of the oscillator. Another example is that of the undamped pendulum, in which, contrary to the harmonic oscillator, the period T depends on the amplitude of oscillation of the pendulum. If we have several different periodic dynamical systems, they are evidently synchronized only if they all have the same period T. Of course, the period being equal does not imply that the different systems are at every time instant in the same dynamical state. For example, in the motion of two harmonic oscillators with the same period, one of the two could have a delayed motion with respect to the other, with the delay being independent of time. Thus, we should speak of synchronization of two or more periodic dynamical elements, either similar or diverse, when they share a common period T. Rather than T, one uses its inverse, the frequency $\nu \equiv 1/T$, or, more often, the pulsation or angular frequency $\omega \equiv 2\pi\nu = 2\pi/T$. The use of ω is so common in the literature on synchronization that by abuse of notation, one usually refers to it as the frequency; we will conform to this usage, assuming no possibility of confusion. Also, for most parts of this book, we will refer to the periodic dynamical elements as oscillators, especially when their specific nature is not relevant. However, it must not be forgotten that these dynamical elements themselves are complex systems, as we emphasize below. The phenomenon of synchronization occurs when a collection of oscillators, which taken individually would have their own natural frequencies, interact with each other and adjust their frequencies so that a large fraction (or, all) of them share a common frequency. When all the oscillators acquire the same frequency, we have complete (or full) synchronization; otherwise, when a sizeable fraction is in synchrony, one has partial synchronization.

Perhaps the first documented observation of synchronization was made by Christiaan Huygens in 1665, about ten years after his invention of the pendulum clock. He observed that two identical pendulum clocks hanging from the same support and set to oscillations differently would soon synchronize perfectly with one another, and in fact, the two would move all the time in opposite directions (when one pendulum reaches the leftmost position, the other would reach the rightmost position). On the face of it, one might argue that being identical, the two clocks are expected to eventually evolve to a state when they would share a common frequency and thus execute a motion in which there is a constant difference among their phases. However, one must realize that however identical they might be, there would be some small and unavoidable construction difference between the two that would result in each clock having its own oscillation frequency. The differences in the frequencies would be sufficiently small so as to guarantee an unprecedented time-keeping precision of each clock, but these would inevitably show up while comparing the oscillations of the two clocks. In the latter case, the phase difference would not be constant in time, so that the two clocks initiated, e.g., in antiphase, would show after a while a different phase relation among them. This is what Huygens himself observed when the two clocks were placed at a distance from one another in the same room. But on being suspended from the same support gave rise to an imperceptible interaction between the two clocks that led to synchronization. A lively description of Huygens' observations is in Strogatz's book [1], where an engrossing and qualitative exposition of several instances of synchronization in physical and biological systems is offered. More examples and a mathematical theory of synchronizing systems are in the fundamental book by Pikovsky, Rosenblum and Kurths [2]; other relevant references are [3–5].

As a matter of fact, it has become evident that synchronization is more pervasive in nature than one might anticipate. To give a set of examples for illustrative purposes, we start from biological systems, mentioning first what is perhaps the most famous example, namely, that of fireflies living in some regions of Southeast Asia that are capable of emitting light signal at regular intervals. It is observed that a swarm of male fireflies gather on trees at night, and, while starting off blinking at individual frequencies, soon begin to flash on and off in perfect synchrony. Another example is the synchronized chirping of crickets. As spectacular as these manifestations might be, one could probably be struck at a deeper level by the realization of how important synchronization is for our own survival. Indeed, the pacemaker cells of the heart must contract in synchrony for its proper functioning. Perhaps the most pervasive rhythmic behavior is the one connected to the circadian rhythm, which is present in the whole of the living world, from plants to simple animals and humans: the adjustment of rhythms of the various functions in an organism is a fascinating example of synchronization. Moving on to physical systems, we may mention the synchronized oscillation of voltage in arrays of Josephson junctions and of the concentration of reagents in a chemical solution. A desired instance of synchronization from everyday life is the one occurring in power-grid networks of distribution of electrical energy, in which generators run in synchrony to be locked in frequency to that of the grid. It is important to realize that, as in the example of the two pendulum clocks of Huygens', the individual elements that synchronize must have in all cases natural frequencies that are distributed in a range of values, even when ideally they should have exactly the same frequency. Such a diversity in natural frequencies is understandably more pronounced in biological systems.

As suggested by the title of this monograph, we will use tools of statistical physics to develop and discuss the mathematical theory of spontaneous synchronization in large population of interacting oscillators. In the framework of the theory of dynamical systems, a great deal has been achieved in the investigation of the behavior of two (or, a few) interacting oscillators. In some cases, when the interaction is the same for every pair of oscillators, analytical results have been obtained also for a relatively large population of oscillators. However, in the general case of many interacting oscillators, a complete analytical characterization of the dynamics is not possible, and one has to resort to statistical tools in order to evaluate the properties of the population. As we go along, we will consider this argument in more detail. Before we proceed, let us briefly summarize the main characteristics of individual oscillators and of the nature of their interaction that can synchronize them.

1.2   The Oscillators and Their Interaction: A Qualitative Discussion

We must now begin to be more precise about the nature of oscillators that can synchronize. It is clear from the foregoing that they are dynamical units, which individually are capable of exhibiting oscillations with a characteristic waveform, amplitude and frequency of oscillation. The latter features depend of course on the physical manifestation of the unit: the heart does not beat the same way as a firefly flashes on and off. Moreover, these characteristic oscillations are such that any (slight) perturbations away from them would soon return the motion to the oscillatory behavior. The dynamics of the individual units should therefore be such as to allow for oscillations that have a characteristic waveform independent of any typical initial condition of the dynamics. The oscillating units should moreover be such that when in interaction with one another, they keep performing a periodic motion, but under suitable conditions may change their frequencies and adjust to a common value. These facts make us conclude that interpreted as dynamical systems, individual oscillators are dissipative, nonlinear systems. We now discuss these aspects.

As mentioned, we have each oscillating unit capable of exhibiting oscillations with a characteristic waveform independent of initial conditions, in the sense that after a transient, the oscillators when starting from any initial state (obviously not completely arbitrary, but included in a given region of the dynamical phase space of the oscillator) will settle into the same motion that has a given frequency. One may then anticipate (correctly) that the dynamics ought to have suitable dissipation and energy-pumping mechanisms so that oscillations that tend to become too large are effectively damped down by dissipation, just as the ones that tend to become too small are suitably pumped up by a supply of energy. Indeed, in the absence of energy supply, a dissipative system would lose its energy, as in the movement of a solid body contrasted by friction. In a pendulum clock, the source of energy is the potential energy of weights that transfer energy to the pendulum through a ratchet-wheel mechanism. In more complex systems, like the biological ones, there might be feedback mechanisms for supply of energy. Due to the two competing tendencies of loss of energy by dissipation and gain through supply of energy, oscillations of a characteristic form, for which pumping and damping effects balance each other, are only sustained. The presence of damping at once precludes the possibility for the underlying dynamics to be conservative, i.e., a dynamics given by the Hamilton equations of motion corresponding to a suitable system Hamiltonian. Consequently, the state that the dynamics relaxes to at long times would be a generic nonequilibrium state [6]. One may recall that the basic tenet of classical equilibrium statistical mechanics is a dynamics due to Hamilton equations of motion derived from a suitable Hamiltonian.

That the dynamics of the individual oscillating units are not derived from an underlying Hamiltonian may also be understood in the following manner. Consider a generic Hamiltonian dynamical system with more than one degree of freedom, and in particular, consider its behavior in the neighborhood of a stable fixed point. The time evolution in the neighborhood of the fixed point, determined by the dynamics linearized around the point, is multiperiodic with frequencies that are in general rationally independent. If nonlinear terms are also taken into account, one has to include harmonics of the basic frequencies, which depend on the initial condition (this is easily seen in, e.g., systems that can be analyzed in action-angle variables). In any case, we should expect a time behavior that is not a simple periodic motion, i.e., characterized by a single frequency. Therefore, we have to preclude our oscillators from being Hamiltonian systems, a conclusion that may be inferred from another consideration. The examples of systems given above clearly refer to quite complex systems. A Hamiltonian description would require their representation as systems with a large number of (microscopic) degrees of freedom. This is almost always a formidable task (think for example to undertake a similar task for a firefly or for a pacemaker cell!). The only hope to represent such systems with equations of motion is to introduce an effective description, with a few effective degrees of freedom and effective interactions. At this level, the equations of motion are not Hamiltonian, and would generically have dissipative terms.

The dynamical properties of the oscillators that we have described above are those pertaining to a limit cycle, and in fact limit cycles often characterize the dynamics of nonlinear dissipative systems. Therefore we have reached the conclusion that the periodic motion of our oscillators should be the limit cycle of a nonlinear dissipative system. Synchronization of a collection of oscillators is a consequence of their interaction. We will refer generally to such a collection as a synchronizing system. As will be shown in more details in the following, the interaction can not be arbitrary, since it must not destroy the oscillatory behavior of the single oscillators.

In the next section, we discuss the characterization of the motion of individual oscillating units in terms of what are known as limit cycles [7]. We start off the section with a brief reminder of the concept of limit cycles.

1.3   Oscillators as Limit Cycles

In order to introduce the concept of limit cycles, we must begin with recalling some
relevant features of dynamical systems. A generic autonomous[1] dynamical system
in $\mathbb{R}^n$ is described by equations of motion of the form

$$\frac{\mathrm{d}x_i}{\mathrm{d}t} = F_i(x_1, x_2, \ldots, x_n; \mu); \quad i = 1, 2, \ldots, n, \tag{1.1}$$

where, typically, one has $n \gg 1$, and we have included a possible dependence of the
functions F_i on a parameter μ. A solution $x_i(t)$, $i = 1, 2, \ldots n$, defines an orbit in
the n-dimensional phase space of the system spanned by the x_i's. For generic F_i's,
such a solution is obtained either through analytic approximation or by performing
numerical integration of (1.1). Let us discuss the properties of two important classes
of possible solutions.

A first class of relevant solutions is given by those for which $x_i(t)$ ∀ i equals
a time-independent constant, i.e., solutions given by the fixed points (also called
equilibrium points) of the dynamics. The latter are defined as points at which all the
F_i's vanish. These points are obviously a function of μ. A fixed point is said to be
stable if an initial condition in a sufficiently small neighborhood of it remains in such
a neighborhood during the dynamical evolution and possibly tends asymptotically
to such a point at long times.[2] In other words, a sufficiently small perturbation of a
stable fixed point remains in its neighborhood during the dynamical evolution.

Another class of solutions, relevant for the scope of this monograph, corresponds
to a periodic motion in the phase space. In this case, we have $x_i(t + T) = x_i(t)$ ∀ i,
where T is the period of the periodic motion. Thus, the orbit is a closed one-
dimensional path in the phase space. For this class of solutions, one has to also
address the issue of stability. It may happen that an orbit starting sufficiently close to
the periodic orbit tends asymptotically to it at long times, in which case the periodic
orbit is said to be stable.[3]

Stable fixed points and stable periodic orbits to which orbits starting nearby tend
asymptotically during their evolution are examples of attractors. The justification

---

[1] A dynamical system is said to be autonomous if the functions F_i do not depend explicitly on time.
Consequently, if $x_i(t)$, $i = 1, 2, \ldots n$, is a particular solution, then also $x_i(t + t_0)$ is a solution for
any choice of t_0, i.e., the origin of time is not relevant. Physically, such a scenario occurs when
one of the following two conditions is realized: either the motion depends only on the interaction
between the dynamical variables of the system and has no external influence, or, even when there
is an external influence, it does not depend explicitly on time.

[2] In Hamiltonian systems, in which time evolution preserves volumes in phase space, the orbit
around a stable fixed point remains close to it but can never tend to it as time progresses. In generic
nonconservative systems, however, both situations may occur; if the orbit tends to the stable fixed
point (which is the more common case), the latter is called asymptotically stable.

[3] Again, the behavior of Hamiltonian systems is different. Conservation of phase-space volumes
prevents the occurrence of orbits that tend to a periodic orbit, whereas one has the common situation of different periodic orbits filling densely the neighborhood of a fixed point. In generic
nonconservative systems, the most common situation is that of an isolated periodic orbit.

behind the name is clear: orbits starting close to a fixed point or a periodic orbit get "attracted" towards the latter as the dynamics proceeds in time. An attracting periodic orbit is what is called a limit cycle, since it is a cycling orbit that is also the limit of orbits starting close to it. In Appendix 1, we show the example of a simple two-dimensional dynamics having a limit cycle as an attractor.

We note that there may be other types of attractors [8]. For example, an attractor could be a subset of the phase space with dimensionality larger than 1, e.g., a torus. Another very interesting and important case is that of the so-called "strange attractors" that occur, e.g., when the dynamics is chaotic, i.e., if it has a positive Lyapunov exponent. Here we are not concerned with these more complex cases, and will instead consider only limit cycles. In Appendix 2, we provide a brief reminder of the definition of Lyapunov exponents.

An orbit starting sufficiently close to a limit cycle will after a while get extremely close to the cycle, although mathematically speaking, it will never reach it due to the uniqueness of solutions of the dynamics. However, from the practical point of view, the orbit will be indistinguishable from the limit cycle. The limit cycles are the periodic motions that characterize the dynamics of the units of synchronizing systems, and their attracting property implies that a small perturbation away from the cycle will soon be damped in time, so that the dynamics will practically coincide with that of the periodic orbit.

Our discussions thus far make us conclude that the periodic motion of the individual oscillators of synchronizing systems should be the stable limit cycle of a dissipative, nonlinear system. Indeed, one can argue quite easily that these cycles can occur only in nonlinear dynamical systems. A linear dynamics $dx_\alpha/dt = \sum_{\alpha,\beta} \Lambda_{\alpha\beta} x_\beta$ can of course generate periodic orbits, but since with every periodic orbit $\{x_\alpha(t)\}$, one may associate a family of periodic orbits $\{cx_\alpha(t)\}$ with c a parameter, such an orbit would not be isolated but be surrounded by an infinite number of periodic orbits obtained by varying c. The issue of which one among the orbits is chosen by the dynamics is set by its initial condition, unlike the independence of the form of a limit cycle with respect to initial conditions. Also, any slight perturbation away from such a closed orbit will unlike a limit cycle not return the motion to the orbit, but will take it to a neighboring closed orbit.

In the case of synchronizing systems, even if the initial conditions of the individual oscillators are not on the limit cycles, but are in their "basin of attraction" (the region in the phase space alluded to above), the oscillators will approach the limit cycles very fast in time, and the subsequent dynamics will be virtually indistinguishable from that on the limit cycles themselves. Since different limit cycles corresponding to different oscillators have in general different oscillation frequency, synchronization among a collection of such oscillators is then possible only as a result of their interaction. The interaction cannot be arbitrary; it must not destroy the oscillatory behavior of the single oscillators, yet give rise possibly to a shift of the frequency so that the oscillators may synchronize. In case of full or partial synchronization, the shift is common to many or all oscillators. The projection of the orbit of the synchronizing system onto the phase space of an individual oscillator is not expected to be very different from the limit cycle of the individual oscillator, a scenario that occurs when the oscillators interact sufficiently weakly. We will see that the latter feature has important consequences on the nature of interaction.

Summarizing, the oscillators of synchronizing systems are nonlinear, dissipative dynamical systems that follow a limit cycle. The frequencies characterizing the limit cycles of the oscillators are distributed within a range of values. The interaction among the oscillators is not strong enough to modify in a substantial way their individual dynamics, but nevertheless can prove to be sufficiently strong to make the oscillators move in unison.

Let us now justify the characterization of the dynamics on a limit cycle in terms of uniform motion of a phase variable, by invoking some simple concepts of dynamical systems theory. To this end, let us consider the periodic orbit representing the limit cycle, and introduce its description with a phase variable. For any solution of the equations of motion (1.1), it is possible to introduce a coordinate given by the length of the curve in the phase space representing the orbit, as

$$s(t) \equiv \int_0^t dt \sqrt{\sum_{i=1}^n \left(\frac{dx_i}{dt}\right)^2}. \tag{1.2}$$

The freedom one has in choosing the origin of time translates to the freedom in the choice of the origin of the curve. For a limit cycle of period T, we have $s(nT) = ns(T)$, or, more generally, $s(nT + t) = s(t) + ns(T)$. As follows from Eq. (1.2), the time rate of variation of s along the path is not generally constant along the path. It proves convenient to parametrize the path in terms of a new variable $\theta \equiv \theta(s)$, such that its rate of variation in time is a constant, which we will identify with the natural frequency $\omega = 2\pi/T$ of the oscillator. Using the definition

$$\theta(s) \equiv \frac{2\pi}{T} \int_0^s \frac{ds'}{\left(\frac{ds}{dt}\right)(s')}, \tag{1.3}$$

one obtains the desired property: $d\theta/dt = (d\theta/ds)(ds/dt) = 2\pi/T = \omega$. Moreover, defining $s_0 \equiv s(T)$, one easily finds that $\theta(s_0) = 2\pi$. Thus, at the end of one period, the value of θ increases by 2π, corresponding to the traversal of a complete orbit. A schematic of the usefulness of the transformation from s to θ is shown in Fig. 1.1.

On the basis of the above discussion, we conclude that a limit-cycle oscillator is completely characterized by a phase θ that changes uniformly in time with period T and frequency ω, according to the equation

$$\frac{d\theta}{dt} = \omega. \tag{1.4}$$

From Eq. (1.4), it follows that the phase θ is a neutrally stable variable; any perturbation to it neither grows nor decays in time.

It turns out that one can use a similar phase description even for orbits that are sufficiently close to the limit cycle, by invoking the concept of *isochrones*, as we now

Fig. 1.1 For a hypothetical limit cycle represented by the closed curve, the figure shows different curve lengths Δs_1 and Δs_2 traversed in the same time interval. By virtue of the transformation (1.3), the phase change nevertheless is the same in both cases

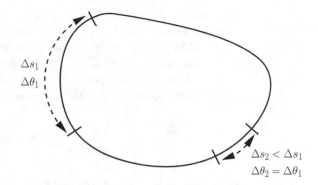

discuss. Consider an initial phase-space point sufficiently close to the limit cycle. The point traverses a path in the phase space, and we consider its successive positions observed at times $t = kT$; $k = 1, 2, 3, \ldots$, where T is the period of the limit cycle. From the properties of the limit cycle, it is evident that the limit of this sequence of points as $k \to \infty$ is a point on the limit cycle, which according to the discussion above has a given value of the phase θ. One then associates this latter value of θ with the sequence of points, which are now said to lie on an isochrone, a $(n-1)$-dimensional hypersurface [2]. In this way, we can associate a phase θ to each point of the phase space lying in the neighborhood of the limit cycle. By this construction, Eq. (1.4) remains valid also in the neighborhood of the limit cycle. Such a construction proves to be particularly useful in discussing the case of many oscillators interacting weakly with one another that we discuss in the next section.

1.4 Interacting Limit-Cycle Oscillators

We now consider the case of two limit-cycle oscillators of frequencies ω_1 and ω_2, which are interacting weakly with one another. We consider the frequencies to satisfy $\min(\omega_1, \omega_2) \gg |\omega_1 - \omega_2|$, a situation typically encountered in synchronizing systems. The governing dynamical equations are

$$\frac{dx_i^{(1)}}{dt} = F_i^{(1)}(x_1^{(1)}, \ldots, x_{n_1}^{(1)}) + \varepsilon G_i^{(1)}(x_1^{(1)}, \ldots, x_{n_1}^{(1)}, x_1^{(2)}, \ldots, x_{n_2}^{(2)}); \quad i = 1, 2, \ldots, n_1,$$

$$\frac{dx_i^{(2)}}{dt} = F_i^{(2)}(x_1^{(2)}, \ldots, x_{n_2}^{(2)}) + \varepsilon G_i^{(2)}(x_1^{(1)}, \ldots, x_{n_1}^{(1)}, x_1^{(2)}, \ldots, x_{n_2}^{(2)}); \quad i = 1, 2, \ldots, n_2, \quad (1.5)$$

where n_1 and n_2 refer to the dimensionalities of the phase space of the two oscillators, and the functions $G_i^{(1)}$ and $G_i^{(2)}$ describe the interaction between the two oscillators, with ε being the coupling constant that is assumed to be small: $|\varepsilon| \ll 1$. To ease the notation, we have here suppressed any possible dependence of the functions on external parameters. We are interested in the dynamics of the two oscillators close to the respective limit cycles associated with their isolated ($\varepsilon = 0$) dynamics. The smallness of ε guarantees that starting from phase-space points sufficiently close to these limit cycles, subsequent evolution remains close to them even in the presence of interaction. We can therefore exploit the concept of isochrones discussed in the preceding section, and associate such points with values of the phase given by θ_1 and θ_2 on the respective isolated limit-cycles of the two oscillators. The chain rule gives

$$\frac{d\theta_k}{dt} = \sum_{i=1}^{n_k} \frac{\partial \theta_k}{\partial x_i^{(k)}} \frac{dx_i^{(k)}}{dt}; \quad k = 1, 2. \tag{1.6}$$

Using Eq. (1.5) to substitute for $dx_i^{(k)}/dt$, and exploiting the fact that Eq. (1.4) is valid also in the neighborhood of the limit cycle, we get

$$\frac{d\theta_k}{dt} = \omega_k + \varepsilon \sum_{i=1}^{n_k} \frac{\partial \theta_k}{\partial x_i^{(k)}} G_i^{(k)}(x_1^{(1)}, \dots, x_{n_1}^{(1)}, x_1^{(2)}, \dots, x_{n_2}^{(2)}); \quad k = 1, 2. \tag{1.7}$$

To leading order in ε, we can now replace the $x_i^{(k)}$'s on the right hand side with the values on the limit cycle belonging to the isochrones characterized by θ_1 and θ_2. We then arrive at the two equations

$$\frac{d\theta_k}{dt} = \omega_k + \varepsilon Q_k(\theta_1, \theta_2); \quad k = 1, 2. \tag{1.8}$$

We now show that the overwhelming contribution to the functions Q_1 and Q_2 comes from a dependence on the difference of the two phases. Introducing the centre-of-mass coordinates

$$\Theta \equiv \frac{\theta_1 + \theta_2}{2}, \quad \Phi \equiv \theta_1 - \theta_2, \tag{1.9}$$

we rewrite Eq. (1.8) as

$$\frac{d\theta_k}{dt} = \omega_k + \varepsilon Q_k(\Theta, \Phi); \quad k = 1, 2, \tag{1.10}$$

where we may expand Q_k in a Fourier series in Θ as $Q_k(\Theta, \Phi) = \sum_p \tilde{Q}_p(\Phi) e^{ip\Theta}$. Next, averaging Eq. (1.10) over a time τ satisfying $1/|\omega_1 - \omega_2| \gg \tau \gg 1/\min(\omega_1, \omega_2)$ (which is a somewhat stronger condition than assuming that $\min(\omega_1, \omega_2) \gg |\omega_1 - \omega_2|$), we get

$$\left\langle \frac{d\theta_k}{dt} \right\rangle \equiv \frac{1}{\tau} \int_0^\tau dt \frac{d\theta_k}{dt} = \omega_k + \frac{\varepsilon}{\tau} \int_0^\tau dt \sum_p \tilde{Q}_p(\Phi) e^{ip\Theta}; \quad k = 1, 2. \tag{1.11}$$

We may now evaluate the right hand side to leading order in ε, by substituting[4] $\Theta \approx (\omega_1 + \omega_2)t/2$, to get

$$\left\langle \frac{d\theta_k}{dt} \right\rangle = \omega_k + \frac{\varepsilon}{\tau} \int_0^\tau dt \, \widetilde{Q}_0(\Phi); \quad k = 1, 2, \tag{1.12}$$

since the terms with $p \neq 0$ give a vanishing contribution. While the left hand side is by definition the averaged time derivative, the integrand on the right hand side is practically constant over the interval $[0, \tau]$ due to our choice of τ, which is the time window chosen to compute the average. We thus arrive at the time-averaged equation

$$\frac{d\theta_k}{dt} = \omega_k + \varepsilon Q(\theta_1 - \theta_2); \quad k = 1, 2, \tag{1.13}$$

where we have dropped the angular brackets on the left hand side and the tilde and the subscript on the right hand side for brevity of notation.

The above treatment may be generalized to the case of any number $N \geq 2$ of weakly-interacting oscillators, allowing to derive for every pair of oscillators an interaction of a form that depends on their phase difference. In order to do such a derivation, we have to assume as in the above that for every pair (i, j), there exists a time τ satisfying $1/|\omega_i - \omega_j| \gg \tau \gg 1/\min(\omega_i, \omega_j)$. For a large number $N \gg 1$ of interacting oscillators, which will be our case of interest, we will consider the natural frequencies to be extracted from a given distribution $g(\omega)$. In such a case, let $[\omega_{\min}, \omega_{\max}]$ denote the range of values of ω over which $g(\omega)$ is appreciably different from zero. The above derivation holds under the assumptions that this range does not include the origin and that there is a time τ satisfying $1/|\omega_{\max} - \omega_{\min}| \gg \tau \gg 1/\min(|\omega_{\min}|, |\omega_{\max}|)$. We thus have for a system of N interacting oscillators the following equation of motion for the i-th oscillator:

$$\frac{d\theta_i}{dt} = \omega_i + \varepsilon \sum_{j=1}^{N}{}' Q_{ij}(\theta_i - \theta_j); \quad i = 1, 2, \ldots, N, \tag{1.14}$$

where we have accounted for the fact that the function Q for a pair of oscillators may depend on the pair under consideration. Here, the prime on the summation sign is to imply that the term with $j = i$ is excluded in the sum.

To summarize, we have seen in this section that even in presence of interaction, the dynamics of oscillators can be described in terms of their phases only. Physically, this could be possible in view of the fact that while weak perturbations do not appreciably affect the dynamics in a direction transversal to the limit-cycles of the individual oscillators, they nevertheless strongly affect the dynamics of their phases. This is because the motion along the limit cycle is neutrally stable.

---

[4]The integral in the right hand side of Eq. (1.11) is preceded by ε. So, to have a term of order ε, it is sufficient that the integral is evaluated at zeroth order. The interaction term in Eq. (1.8) is of order ε, so at zeroth order Θ is given by $(\omega_1 + \omega_2)t/2$.

In this monograph, we will exclusively consider the case in which every oscillator in a collection of N interacting oscillators is coupled equally to every other. As a result, one may perform a change of notation $f \equiv \varepsilon Q_{ij}$ to write (1.14) as

$$\frac{d\theta_i}{dt} = \omega_i + \frac{K}{N} \sum_{j=1}^{N}{}' f(\theta_i - \theta_j), \tag{1.15}$$

where we have also introduced a common coupling constant K and a prefactor of $1/N$ in the second term on the right hand side. In this way, the strength of the interaction between the oscillators is normalized with their total number N. This normalizing procedure, called "Kac's prescription", allows to have a well-defined limit of the associated term as $N \to \infty$ [9]. The reader may object that this procedure sounds unphysical, since the strength of interaction between any pair of oscillators is not expected to change with the number of oscillators. Although this objection is perfectly justified, nevertheless the procedure is convenient and useful to obtain results pertaining to a system where this normalization is not introduced. In fact, as will be clear in the next chapters, with this normalization one obtains at the point of transition between synchronized and unsynchronized states relations between the parameters of the system (e.g., the width of the frequency distribution and the coupling constant K) that do not contain N, as long as N is very large. In other words, the transition occurs at points of the so-called thermodynamic phase space that are independent of N. The transition points for a system where the $1/N$ normalization has not been done can then be simply obtained by scaling back, by using the concrete value of N under consideration.

Equation (1.15) would from now on be our main equations of motion for the study of interacting oscillators. The celebrated Kuramoto model, which is a proto-type model for a system of interacting oscillators exhibiting spontaneous synchro-nization [10–15], is obtained from Eq. (1.15) with the particular choice $f(\theta_i - \theta_j) = -\sin(\theta_i - \theta_j)$. The Kuramoto equations of motion are thus

$$\frac{d\theta_i}{dt} = \omega_i - \frac{K}{N} \sum_{j=1}^{N} \sin(\theta_i - \theta_j); \quad i = 1, 2, \ldots, N, \tag{1.16}$$

where the frequencies ω_i's for the different oscillators are extracted independently from a common distribution $g(\omega)$ with the normalization $\int_{-\infty}^{\infty} d\omega \, g(\omega) = 1$. Note that the prime in the sum appearing in Eq. (1.15), implying that the term $j = i$ is not to be included in the sum, need not be put explicitly in Eq. (1.16) as the sine interaction automatically takes care of this constraint.

1.5 Synchronizing Systems as Statistical Mechanics Systems

Synchronization in a large number of oscillators in interaction is a remarkable example of an emergent behavior. Since the natural frequencies of the oscillators are different, the dynamical attainment of a common frequency by all or a large portion of them may be compared to the spontaneous attainment of common orientation of interacting spins in a magnetic system. In fact, we will see that in simple models of synchronizing systems, e.g., the Kuramoto model, one introduces an order parameter for the amount of synchronization that is quite similar to the magnetization of magnetic systems.

As we have anticipated, an exact analytical characterization of the dynamics of a large number of interacting oscillators is not quite possible in general. A fortiori this is true when we consider the stochastic fluctuations that are present in the oscillators, which can affect their dynamical parameters. We remind that we are going to use an effective description of real concrete synchronizing systems; a few degrees of freedom represent each oscillator, which in reality is quite a complex system. Introducing stochastic fluctuations in the parameters of the effective description, in particular, of the natural frequency of the oscillator, takes somewhat into account the complexity of the real oscillator. Moreover, the synchronizing system should often be considered as not isolated, but subject to external noise. The introduction of stochastic fluctuations in the models can represent in an effective way all sources of variability.

The study of many-body systems subject to stochastic fluctuations is the common ground of statistical physics. Let us establish the framework within which synchronizing systems are treated. We do this in steps, beginning with the case of a single oscillator. The following is a short review of theoretical tools that are probably already known to most readers. However, since we will use them in most of the analysis contained in subsequent chapters, we find it useful to give a self-contained presentation.

1.5.1 A Single Oscillator with Noise

We have already seen that the oscillators constituting a synchronizing system can be effectively described with only one degree of freedom, i.e., the phase θ. Let us consider a single such oscillator evolving according to Eq. (1.4). Although the trivial solution of this equation,

$$\theta(t) = \theta(0) + \omega t, \tag{1.17}$$

gives an unbounded increase of $\theta(t)$ (or, of $|\theta(t)|$ if ω is negative), one has to consider that physically the phase is defined modulus 2π.

We now introduce noise in the dynamics, by adding to the equation of motion (1.4) a stochastic term, thus obtaining the Langevin equation

$$\frac{d\theta}{dt} = \omega + \eta(t). \tag{1.18}$$

The noise $\eta(t)$ is a stochastic variable with its average equal to 0 and having delta correlation in time:

$$\langle \eta(t) \rangle = 0, \quad \langle \eta(t)\eta(t') \rangle = 2D\delta(t - t'), \tag{1.19}$$

with D determining the intensity of the noise, and angular brackets denoting averages over noise realizations. The Langevin equation (1.18) is trivially integrated to yield

$$\theta(t) = \theta(0) + \omega t + \int_0^t dt' \, \eta(t'). \tag{1.20}$$

By averaging over noise realizations, we obtain the diffusive behavior of the phase with respect to its perfectly periodic behavior (1.17), as

$$\langle (\theta(t) - \theta(0) - \omega t)^2 \rangle = 2Dt. \tag{1.21}$$

The constant D is therefore called the diffusion coefficient. As is well known, the physical meaning of the average over noise realizations is the computation of the average of the observable between angular brackets (in this case, the mean-squared displacement with respect to the perfectly periodic behavior) over many realizations of the same dynamics. If the noise is not delta correlated as in the second equation in (1.19), but $\langle \eta(0)\eta(t) \rangle$ is a smooth function of time, then D can be defined as half of the integral, from $t = -\infty$ to $t = +\infty$, of the correlation function $\langle \eta(0)\eta(t) \rangle$. Then, for t large with respect to the decay time of the correlation, the phase diffusion is still given by Eq. (1.21).

If the oscillator is subject to an external field $F(\theta)$, the equation of motion without noise becomes

$$\frac{d\theta}{dt} = \omega + F(\theta). \tag{1.22}$$

With a slight abuse of notation, we will refer to the field as a force. In a synchronizing system, the external field will be the result of interactions with other oscillators (and would therefore depend also on their phases). Provided they exist, the values of θ for which $\omega + F(\theta) = 0$ denote fixed or equilibrium points of the oscillator. An equilibrium point at θ_0 is stable if the derivative of $F(\theta)$ is negative at θ_0. On the other hand, if $\omega + F(\theta)$ does not vanish for any value of θ, then the phase increases (or decreases) monotonically, although not at a uniform rate. With the inclusion of noise, we have the Langevin equation in presence of $F(\theta)$, as

$$\frac{d\theta}{dt} = \omega + F(\theta) + \eta(t).$$ (1.23)

Although in many cases, depending on the functional form of $F(\theta)$, this equation can be integrated, some qualitative features can be inferred without solving it. If $\omega + F(\theta)$ is of definite sign, the noise adds diffusion to the nonuniform monotonic variation of the phase, quite similarly to the case without external force. On the other hand, if there are equilibrium positions in the noiseless case, the noise would not allow the oscillator to remain in them, and we will have a diffusive behavior of θ determined both by the size of D and by the shape of $F(\theta)$ around the equilibrium point. A clarifying example is presented in Sect. 1.5.1.1.

The statistical description of the Langevin equation (1.23) is obtained by introducing the distribution function $\rho(\theta, t)$, whose meaning is the following: if we have a large number of oscillators, each one performing a realization of the dynamics (1.23), the quantity $\rho(\theta, t)d\theta$ gives the fraction of oscillators at time t that are found to have phases between θ and $\theta + d\theta$. The evolution of the distribution function $\rho(\theta, t)$ obeys a partial differential equation, i.e., a Fokker-Planck equation given by

$$\frac{\partial \rho}{\partial t} = -\frac{\partial}{\partial \theta}\left[(\omega + F(\theta))\,\rho\right] + D\frac{\partial^2 \rho}{\partial \theta^2}.$$ (1.24)

The passage from the Langevin equation of motion to the Fokker-Planck equation for the distribution function is described in detail in textbooks dedicated to stochastic differential equations (see, e.g., [16]). It is shown that in the case of a delta-correlated noise, there is no approximation in this passage if the noise has a Gaussian distribution; we will therefore assume here that the noise has a Gaussian distribution.[5]

The Fokker-Planck equation can be written in the form

$$\frac{\partial \rho}{\partial t} = L_{FP}(\theta)\rho,$$ (1.25)

where $L_{FP}(\theta)$ is the linear differential Fokker-Planck operator

$$L_{FP}(\theta) \equiv -\frac{\partial}{\partial \theta}(\omega + F(\theta)) + D\frac{\partial^2}{\partial \theta^2}.$$ (1.26)

---

[5]For other noise distributions, there are in generally additional terms with higher order derivatives of $\rho(\theta, t)$ with respect to θ, resulting in a partial differential equation of higher order than the Fokker-Planck equation. However, on physical grounds we can expect that in the more general case, the picture would not be different, apart from small quantitative differences. As a matter of fact, the Fokker-Planck equation is by far the most used tool in the study of stochastic phenomena. Actually, if the noise distribution has compact support, i.e., if the noise is strictly bounded, then, e.g., the escape from a potential well, which with a Gaussian noise has always a probability different from zero (although it could be extremely small), might be impossible, depending on the maximum size of the noise, the height of the potential well and its slope. We do not consider this subtle point here, and stick to the usual case of Gaussian noise.

Given the distribution $\rho(\theta, 0)$ at time $t = 0$, the formal solution of the Fokker-Planck equation is

$$\rho(\theta, t) = \exp[L_{FP}(\theta)t]\,\rho(\theta, 0). \tag{1.27}$$

Solving this equation is not any simpler than solving the Langevin equation (1.23). However, we are mainly interested in stationary solutions, which from Eq. (1.25) are obtained by solving

$$L_{FP}(\theta)\rho = 0. \tag{1.28}$$

We expect that a stationary solution gives the distribution function at sufficiently long times, i.e., the distribution one has after a transient. This expectation is justified by the fact that under mild conditions on the function(s) appearing in the Fokker-Planck operator (in our case, we have just the single function $F(\theta)$), the stationary solution is unique, and at long times, any solution (identified by the initial condition $\rho(\theta, 0)$) approaches the stationary solution [16].

1.5.1.1 The Stationary States in a Representative Case

A stationary solution of the Fokker-Planck equation is said to represent a stationary state of the system. Here we discuss the stationary state in a particular case. We will not show the details of the computation, which although simple will be deferred to the next chapter. Our purpose in this section is just to give a flavor of the role of the various quantities appearing in the equation.

The external force $F(\theta)$ is a 2π-periodic function. For our example, we choose the simple function $F(\theta) = -\sin\theta$: it represents a force that attracts the oscillator phase to $\theta = 0$. For definiteness, we assume that ω is positive. In the noiseless case, we see from Eq. (1.22) that the equilibrium points, when they exist, are the points θ_0 for which $\sin\theta_0 = \omega$. Clearly, for $\omega > 0$, such values of θ_0 exist only if $\omega \le 1$; in that case, there are two solutions for θ_0 (for ω strictly smaller than 1): one between $\theta = 0$ and $\theta = \pi/2$, and one between $\theta = \pi/2$ and $\theta = \pi$. However, only the former is stable. On the other hand, if $\omega > 1$, there is no equilibrium point, and θ increases monotonically with time at a nonuniform rate, which is minimum for $\theta = \pi/2$ and maximum for $\theta = 3\pi/2$. Going now to the case with noise, we need to consider the Fokker-Planck equation

$$\frac{\partial \rho}{\partial t} = -\frac{\partial}{\partial \theta}\left[(\omega - \sin\theta)\,\rho\right] + D\frac{\partial^2 \rho}{\partial \theta^2}. \tag{1.29}$$

The stationary state is obtained by solving the ordinary differential equation that one gets by equating to zero the right hand side. The explicit solution will be given in Chap. 2 while treating synchronizing systems. Our aim here is only to show the plots

Fig. 1.2 Stationary distribution of the Fokker-Planck equation (1.29) for $\omega = 1.2$ and three different values of the diffusion coefficient D: $D = 0$ (full line), $D = 0.3$ (dot-dashed line), and $D = 2$ (dashed line)

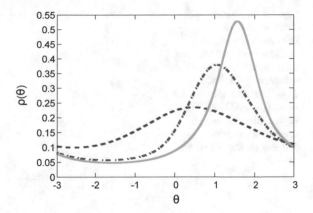

of the stationary state for some values of the natural frequency ω and the diffusion coefficient.[6]

Let us begin with the case $\omega > 1$. In Fig. 1.2, we show the stationary distribution for $\omega = 1.2$ and three different values of the diffusion coefficient D, i.e., $D = 0$, $D = 0.3$ and $D = 2.0$. The first value refers to the noiseless situation; since there is no equilibrium position and θ increases monotonically with time, it is no surprise that the equilibrium distribution spreads all over $[0, 2\pi]$, see the full line in Fig. 1.2. The probability to find the oscillator is maximum for $\theta = \pi/2$, when the rate of variation of θ is minimum. Introducing the noise with $D > 0$ flattens the distribution, as can be seen from the other two curves in Fig. 1.2. It is interesting to note that the maximum of the distribution varies with D, and that it does not remain in the position it had for $D = 0$.

We now turn our attention to the case $\omega < 1$. In Fig. 1.3, we plot the stationary distribution for $\omega = 0.5$ and two different values of the diffusion coefficient, $D = 0.3$ and $D = 2$. Since the noiseless case has now an equilibrium position for $\theta = \pi/6$, the stationary distribution in this case is a delta function centered at this value of θ; in the plot, this is indicated with a vertical line at the corresponding position. We see that even a very small amount of noise is sufficient to have a stationary distribution that is different from zero in the whole range of values of the phase; obviously, for very small D, the distribution will be sharply peaked. Also in this case, we note that the maximum of the distribution varies with D.

---

[6] We do not lose generality by assuming that the coefficient of the sinusoidal external force is unity; in fact, if it had a given value $a > 0$, the stationary state would depend on the ratios ω/a and D/a. The assumption $a = 1$ could be made also in the study of the transient behavior (i.e., of the time evolution of $\rho(\theta, t)$ towards the stationary state), by simply rescaling the time.

Fig. 1.3 Stationary
distribution of the
Fokker-Planck equation
(1.29) for $\omega = 0.5$ and two
different values of the
diffusion coefficient D:
$D = 0.3$ (full line), and
$D = 2$ (dashed line). The
vertical line at $\theta = \pi/6$
represents the delta
distribution that occurs for
$D = 0$

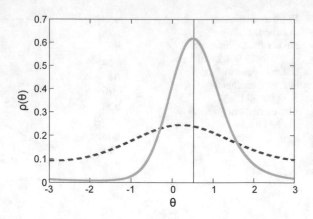

1.5.2 *Oscillators in Interaction*

It is with the study of interacting oscillators that the usefulness of the statistical
approach becomes more evident. In this case, the force acting on each oscillator
is quite generally due to its interaction with all the other oscillators. The coupled
Langevin equations for the noisy dynamics of a system of N interacting oscillators,
assuming for the moment that the natural frequency is the same for each, are

$$\frac{d\theta_i}{dt} = \omega + \frac{K}{N} \sum_{j=1}^{N}{}' f(\theta_i - \theta_j) + \eta_i(t) \,, \qquad i = 1, 2, \ldots, N, \qquad (1.30)$$

where η_i is the stochastic noise acting on the i-th oscillator. One usually considers
$\eta_i(t)$ to be a Gaussian, white noise, chosen independently for each oscillator, so that
similar to Eq. (1.19), one has

$$\langle \eta_i(t) \rangle = 0, \quad \langle \eta_i(t)\eta_j(t') \rangle = 2D\delta_{ij}\delta(t - t'). \qquad (1.31)$$

Similar to the single-oscillator case, one may in the present situation obtain under
some conditions a Fokker-Planck equation for the distribution function $\rho(\theta, t)$. As
we have remarked, the distribution function $\rho(\theta, t)$ describes for the single oscillator the outcome of many realizations of the dynamics (1.23): since the noise is a
stochastic variable, the distribution of the phases of the oscillators (each one following one realization of the dynamics) at time t is given by $\rho(\theta, t)$. The interpretation
of the distribution function is the same for a system of interacting oscillators, i.e.,
it is the average over many realizations of the dynamics. However, especially when
confronted with the results of numerical simulations, we always make a comparison
between $\rho(\theta, t)$ as obtained by solving the Fokker-Planck equation and the average over the oscillators of the system in the simulation of a single realization of the
dynamics. The conceptual difference should be clear, although it is often overlooked.

In Appendix 3, we make some more comments concerning this issue, which are in general also relevant to the study of many-body systems.

In the present situation, it may be shown that one obtains an equation of the same form as Eq. (1.24), but where the force $F(\theta)$ is no more an external force, but rather a time-dependent function given by

$$F(\theta, t) = \int_0^{2\pi} d\theta' \, f(\theta - \theta')\rho(\theta', t). \tag{1.32}$$

We thus have the corresponding Fokker-Planck equation

$$\frac{\partial\rho}{\partial t} = -\frac{\partial}{\partial\theta}\left[(\omega + F(\theta, t))\,\rho\right] + D\frac{\partial^2\rho}{\partial\theta^2}. \tag{1.33}$$

The derivation of this equation (and of its generalization shown just in the next paragraph) from an N-body Fokker-Planck equation will be given and justified in Chap. 2. We note an important difference of Eq. (1.33) with respect to the one for a single oscillator subject to an external force given by Eq. (1.24): in the former, the distribution $\rho(\theta, t)$ appears implicitly in the force term $F(\theta, t)$, so that the equation is no more linear. This fact has another consequence: The conditions of the theorem that asserts the uniqueness of the stationary state, together with the property that any initial distribution $\rho(\theta, 0)$ converges at long times to such stationary distribution, are no longer verified. In fact, one condition is that the drift coefficient, i.e., the expression multiplying ρ in the first term on the right hand side of the Fokker-Planck equation (1.33), is a given function of θ and of time t; in our case, it is a function of ρ itself, so that, e.g., two different solutions have at a given time t different drift coefficients. However, it is expected that at long times, the distribution $\rho(\theta, t)$ will nevertheless converge to a stationary distribution (if and when it exists), and Eq. (1.33) is the key tool for the study of interacting oscillators that are described only by their phases.[7]

In contrast to the case of the same natural frequency of all the oscillators treated above, we will more generally be concerned with synchronizing systems in which the natural frequencies are different among the oscillators, i.e., with the frequencies distributed according to a given distribution $g(\omega)$. Treating such a case requires a generalization of Eqs. (1.32) and (1.33). In fact, the distribution functions will now depend also on ω, or, in other words, we need a distribution for each value of ω included in the support of $g(\omega)$. The Fokker-Planck equation will have the same form as in Eq. (1.33), which now reads

$$\frac{\partial\rho(\theta, \omega, t)}{\partial t} = -\frac{\partial}{\partial\theta}\left[(\omega + F(\theta, t))\,\rho(\theta, \omega, t)\right] + D\frac{\partial^2\rho(\theta, \omega, t)}{\partial\theta^2}. \tag{1.34}$$

---

[7] Actually, we will see in the next Chapter that a stationary distribution representing a synchronized state exists only after making a simple time-dependent change of variables.

Mathematically, one has a system of such equations for each value of ω that are coupled through the force term

$$F(\theta, t) = \int d\omega \int_0^{2\pi} d\theta'\, g(\omega) f(\theta, \theta') \rho(\theta', \omega, t). \qquad (1.35)$$

We remind that the frequency distribution $g(\omega)$ is normalized:

$$\int d\omega\, g(\omega) = 1. \qquad (1.36)$$

1.6 The Features of a Statistical Physics Description

The Fokker-Planck equation is the principal tool of choice for theoretical study of a collection of oscillators in interaction in the presence of noise. In fact, it appears more manageable than the system of Langevin equations (1.30). Its main utility is probably in the determination of stationary distributions $\rho(\theta, \omega)$, which are obtained by setting the left hand side of Eq. (1.34) to zero. In this case, also the force term $F(\theta, t)$ given by Eq. (1.35) is time independent, just as $\rho(\theta, \omega, t)$ will be.

Although, as mentioned, there is no theorem guaranteeing the approach to a stationary state, nevertheless this occurs in most situations. It is in the stationary states that we look for the occurrence of synchronization. Let us see qualitatively what we expect when the system of oscillators synchronizes and when it does not. For this, we assume that the interaction $f(\theta_1 - \theta_2)$ (see Eq. 1.30) is such that it tends to equalize their phases. We can be guided by the example above in which a single oscillator with natural frequency ω is subject to an external force equal to $-\sin\theta$, which tends to make the phase equal to 0. We see from Figs. 1.2 and 1.3 that as a result of the interplay between the driving due to ω and the external force, a stationary state develops, in which some regions of θ are more probable than others. The difference in probability depends on the level of noise.

We begin by considering the simple case of two oscillators with different natural frequencies ω_1 and ω_2, with an interaction between them that tends to make their phases equal. For definiteness, we assume $\omega_1 > \omega_2 > 0$ and $f_1(\theta_1, \theta_2) = -\sin(\theta_1 - \theta_2) = -f_2(\theta_1, \theta_2)$, where f_i is the force on the i-th oscillator due to interaction. In absence of noise, we have the equations of motion

$$\frac{d\theta_1}{dt} = \omega_1 - \sin(\theta_1 - \theta_2), \quad \frac{d\theta_2}{dt} = \omega_2 + \sin(\theta_1 - \theta_2). \qquad (1.37)$$

With the change of variables $\theta_T \equiv \theta_1 + \theta_2$ and $\phi \equiv \theta_1 - \theta_2$, we obtain the two uncoupled equations

$$\frac{d\theta_T}{dt} = \omega_1 + \omega_2, \quad \frac{d\phi}{dt} = \Delta\omega - 2\sin(\phi), \qquad (1.38)$$

with $\Delta\omega \equiv \omega_1 - \omega_2 > 0$. The first equation shows that the "center of mass" of the two oscillators moves at a uniform rate; the second equation is the same as Eq. (1.22) with the substitutions $\theta \to \phi$ and $F(\theta) \to -2\sin\phi$. From the analysis in Sect. 1.5.1.1, we know that in the stationary state, ϕ has a definite value if $\Delta\omega < 2$, while it has a distribution in $[0, 2\pi]$ if $\Delta\omega > 2$. In the first case, we see that the two oscillators keep a constant phase difference equal to the definite value of ϕ, and that they move according to Eq. (1.38) with the same frequency $\omega_0 \equiv (\omega_1 + \omega_2)/2$. This "frequency locking" and the consequent "phase locking" (i.e., the constant phase difference between the two oscillators) are trademarks of synchronization. Because of interaction, the oscillators adjust and equalize their natural frequencies, the one with the larger one adjusting by decreasing it, the other by increasing it. In the case $\Delta\omega > 2$, the locking does not occur, although ϕ does have a larger probability for certain values, while without interaction, the distribution of ϕ would be uniform. If the oscillators are noisy, with noise terms $\eta_1(t)$ and $\eta_2(t)$, Eq. (1.38) is replaced by

$$\frac{d\theta_T}{dt} = \omega_1 + \omega_2 + \eta_T(t), \quad \frac{d\phi}{dt} = \Delta\omega - 2\sin(\phi) + \Delta\eta(t), \tag{1.39}$$

where $\eta_T(t) = \eta_1(t) + \eta_2(t)$ and $\Delta\eta(t) = \eta_1(t) - \eta_2(t)$, with $\langle\eta_T(t)\eta_T(t')\rangle = \langle\Delta\eta(t)\Delta\eta(t')\rangle = 4D\delta(t - t')$. Again, from Sect. 1.5.1.1, we see that with the noise, there is never a real "frequency locking", (or "phase locking"), since the phase difference ϕ has a distribution in the stationary state that is different from 0 in the whole of the range $[0, 2\pi]$. The physical reason behind this can be understood easily. In the presence of noise, we cannot even for a single oscillator with a constant natural frequency speak of a uniform variation of the phase; the fluctuations and the diffusion caused by the noise transform the natural frequency to an average frequency, i.e., the natural frequency itself has fluctuations (we remind that the noise has been introduced primarily just to reproduce these fluctuations common in concrete systems [17]). If the natural frequencies of two interacting oscillators have fluctuations, we expect that also the frequency associated to their phase difference ϕ has fluctuations. Therefore, the locking concept is in a sense smoothed: we can talk of synchronization when the average frequency of the phase difference (average with respect to the noise distribution) is very small in absolute value; then, the phases θ_1 and θ_2 will be for most of the times "locked", with jumps from time to time. This approximate locking will be more and more pronounced the smaller the noise and the smaller $\Delta\omega$ is; for an intensity of the force different from unity, the relevant quantities will be the ratios of $\Delta\omega$ and of D with the force intensity.

We are now in a position to understand qualitatively what can happen when there are many oscillators in interaction. We assume that the force between any pair of oscillators tends to equalize their phases. Suppose we have a configuration at a given time in which the oscillators are distributed more or less uniformly on $[0, 2\pi]$; then, it is easy to see that the total force acting on each oscillator will be small. On the other hand, if a large proportion of oscillators happen to have quite similar phases, then the rest of the oscillators will experience a force tending to pull their phase towards that of the group. If the number of oscillators with nearly equal phases increases, the force on the remaining oscillators will be even stronger, favoring further grouping. The dynamics might cause an alternation between the two situations, but we can expect that on increasing the intensity of the pair force, the tendency to group increases. Then, by increasing the pair force intensity, we expect to arrive at a situation of phase locking, although in the approximate sense explained above for the presence of noise. In Chap. 2, we will quantitatively formalize these issues.

1.6.1 The Advantages of the Fokker-Planck Equation

As we have already remarked, the main use of the Fokker-Planck equation (1.34) is in the determination of the stationary states $\rho(\theta, \omega)$. To obtain the same information from the equations of motion, one would have to obtain numerically the solution of the system of coupled Langevin equations (1.30) (where now ω would be replaced with ω_i, since each oscillator would have a natural frequency extracted from the given distribution $g(\omega)$), and study the long time behavior of this solution. This solution is what is actually obtained in numerical simulations (see Appendix 3 at the end of the chapter for the interpretation of the comparison between theory and numerical simulations); however, it is conceptually satisfying to have an analytical theory that allows to obtain an evaluation of this long-time behavior.

From the operative point of view, the passage from the coupled Langevin equations to the Fokker-Planck equation can be seen as the passage from the study of an N-body dynamics to a single-particle dynamics, since the distribution $\rho(\theta, \omega, t)$ depends on a single variable θ. The price that we pay for this is the passage from ordinary differential equations to partial differential equations; however, while the former are stochastic, the latter are deterministic, with the diffusion due to the noise accounted for by the second-order derivative with respect to θ. It should also be stressed that once a stationary solution of the Fokker-Planck equation has been obtained, one has to study its stability with respect to perturbations, since only stable stationary solutions are physically acceptable representation of stationary states. The study of the stability of a stationary solution is not always easy, although in important cases, e.g., for homogeneous solutions, the analysis is relatively simple.

The force term as given by Eq. (1.35) is rightfully termed a mean-field force, since it is the average force on an oscillator due to the distribution of phases of all the oscillators. It is not obvious that a mean-field force is a good approximation for a closed equation for the one-particle distribution function. As a matter of fact, very often the approximation of a mean-field term is a bad approximation, and one has to resort to other approximate closed equations for the single-particle distribution function; perhaps the most widely known equation of this type is the Boltzmann equation, which is suitable for dilute gases. However, there is a case in which the mean-field approximation is a very good one, namely, the case of long-range interactions. The interaction generally adopted in synchronizing systems of oscillators, with any pair of oscillators interacting with the same force expression, falls in this case. This will be exploited in the derivation of the Fokker-Planck equation from the N-body Fokker-Planck equation in the next chapter, assuming $N \gg 1$. Although the approximation worsens if N is not very large, or, if the interaction between the oscillators cannot be considered as long-range, the Fokker-Planck equation is still a good starting point for the study of the behavior of the system.[8]

The evaluation of stable stationary distributions as a function of the parameters (e.g., the noise intensity, the intensity of the interaction, the width of the frequency distribution $g(\omega)$) allows to find the values of these parameters for which the behavior changes qualitatively; these relevant points in the parameter space denote phase transitions. The passage from a synchronized state to a non-synchronized state can be viewed as a phase transition in a statistical system.

The reader may wonder why up to now we have always talked about the force and have never discussed the aspect of a potential energy; after all, the force including the natural frequency term, given by $\omega - \sin\theta$, can be derived from a potential energy equal to $-(\omega\theta + \cos\theta)$. The problem with this approach is that such a potential is not 2π periodic, which has important consequences. We know that the stationary state of the Fokker-Planck equation (1.24) when the force term $\omega + F(\theta)$ is replaced with minus the derivative of a 2π-periodic potential, i.e., with $-(\partial U(\theta)) / (\partial\theta)$ with $U(\theta + 2\pi) = U(\theta)$, is simply given up to a normalizing factor by $\exp[-U(\theta)/D]$. This is no more the case if such a periodic $U(\theta)$ does not exist, which in our case is due to the presence of the natural frequency ω. This fact makes the search for the stationary states more difficult than for systems with physically meaningful potential energy. This has also the physical consequence that the stationary states cannot be classified as thermodynamic equilibrium states.[9]

Finally, we note that the case of noiseless systems can be obtained in the limit of vanishing noise. Mathematically, this may not be a well defined procedure, since the degree of the Fokker-Planck equation changes for $D = 0$. However, it is found in most cases that the limit reproduces numerical and analytical results. Concerning this point, we want to discuss in the next section some results that have been obtained for the solution of the equations of motion in the noiseless case.

1.7   Some Results for Noiseless Interacting Oscillators

In this section, we present two results concerning the noiseless Kuramoto model, Eq. (1.16). In the remaining of this section, we are going to study two different problems related to the equations of motion (1.16): (i) the long-time behavior of the system of oscillators for a unimodal distribution $g(\omega)$ that is symmetric about the average frequency $\omega_0 = \int d\omega \, \omega g(\omega)$; (ii) the dynamical evolution determined by Eq. (1.16) for a particular distribution $g(\omega)$. Since the number N of oscillators is

---

[8]This can be paralleled with the elementary study of magnetic systems: although magnetic units generally interact only at distances on atomic scale, a mean-field approximation can be employed to obtain easily their qualitative behavior.

[9]This in turn is due to the absence of detailed balance (see Appendix 1 of Chap. 3).

typically very large, which in principle could be even infinite, use is made of the distribution function $\rho(\theta, \omega, t)$, but the treatment of this noiseless case does not use concepts of statistical physics.

1.7.1 The Kuramoto Solution

Let us suppose that the distribution $g(\omega)$ satisfies $g(\omega_0 + \omega) = g(\omega_0 - \omega)$, and that for any pair (ω_1, ω_2) such that $0 \leq \omega_1 \leq \omega_2$, we have $g(\omega_0 + \omega_1) \geq g(\omega_0 + \omega_2)$. These relations imply that $g(\omega)$ is symmetric about the average frequency ω_0 and that it is unimodal, i.e., it is not increasing for $\omega \geq \omega_0$ (and thus not decreasing for $\omega \leq \omega_0$).

As a first step of our analysis, we make a transformation of coordinates that corresponds to a passage in the frame of observation, from the laboratory frame to a frame of reference that rotates uniformly at frequency ω_0 with respect to the laboratory frame: $\theta_i = \theta_i' + \omega_0 t$ for each i. Defining $\omega_i' \equiv \omega_i - \omega_0$, the equations of motion with the new primed variables are identical to Eq. (1.16), but now the distribution $g(\omega')$ is symmetric about the average frequency $\omega = 0$, having been obtained by shifting $g(\omega)$ by ω_0. From now on, we remove the primes in the variables in order not to overload the notation.

As a second step, it is convenient to think of the oscillator phases as a collection of points moving on a unit circle. Then, at any time t, one may associate a vector of unit length to each point, take a vector sum, and divide by N, to get a vector of length $r(t)$ inclined at an angle $\psi(t)$ with respect to a reference axis [11, 12]:

$$r(t)e^{i\psi(t)} \equiv \frac{1}{N} \sum_{j=1}^{N} e^{i\theta_j(t)}. \tag{1.40}$$

Here, $\psi(t)$ gives the average angle, while $r(t)$ measures the amount of phase coherence or synchrony in the system at time t (thus serving the role of an order parameter). Indeed, if the phases are scattered around randomly on the circle, one has $r(t) = 0$, while, by contrast, if the oscillator phases are clustered together on the circle, we have $r(t) > 0$. In the extreme case when all the oscillator phases have the same value, $r(t)$ attains its maximum possible value of unity.

Multiplying both sides of Eq. (1.40) by $e^{-i\theta_i}$ and taking the imaginary part, we obtain the equality

$$r \sin(\psi - \theta_i) = \frac{1}{N} \sum_{j=1}^{N} \sin(\theta_j - \theta_i). \tag{1.41}$$

Substituting in the equations of motion (1.16), we obtain

$$\frac{d\theta_i}{dt} = \omega_i - Kr \sin(\theta_i - \psi); \quad i = 1, 2, \ldots, N. \tag{1.42}$$

These equations look as though the N oscillators are uncoupled, but actually of course they are coupled through r and ψ that depend on all the θ's. Nevertheless, the above form is useful for analytical treatment as it evidently illustrates that during the dynamical evolution, each oscillator is attracted towards the average phase ψ, with a strength that for a given K is proportional to the modulus r of the order parameter.

Both the modulus and the phase of the order parameter depend on time through the time dependence of the phases of the oscillators. In his analysis of the model [10, 11], Kuramoto argued that the dynamics of the oscillators would settle at long times into a state in which r and ψ become time independent. Correspondingly, one would have the phase of the order parameter rotating uniformly in the laboratory frame of reference at the frequency ω_0. In other words, one would have a stationary state in the rotating frame. By redefining, if necessary, the origin of the phases, we may set ψ to zero. Therefore, the stationary-state dynamics reads

$$\frac{d\theta_i}{dt} = \omega_i - Kr \sin\theta_i; \quad i = 1, 2, \ldots, N, \tag{1.43}$$

where the quantity r in the above equation is to be considered time independent. We have already encountered the problem represented by Eq. (1.43) in Sect. 1.5.1.1, although in that example, the term proportional to $-\sin\theta$ was due to an external field. We know that there are two possible behaviors depending on the relative magnitude of ω_i and Kr. For $|\omega_i| < Kr$, the phase θ_i reaches a stable fixed point: of the two fixed points given by $\theta(\omega_i) = \arcsin(\omega_i/Kr)$; $\cos\theta(\omega_i) = \pm\sqrt{1 - \omega_i^2 K^2/r^2}$, it is easily seen that the stable fixed point is the one with a positive cosine.[10] We thus have the stable fixed-point value $\theta(\omega_i) = \arcsin(\omega_i/Kr)$; $\cos\theta(\omega_i) = \sqrt{1 - \omega_i^2 K^2/r^2}$. For $|\omega_i| > Kr$, Eq. (1.43) does not admit a fixed point, but nevertheless, a stationary state for the systems of oscillators can be found by invoking the following argument.

First, Kuramoto invokes at this stage the use of the distribution $\rho(\theta, \omega)$ to characterize the stationary state by considering the system of oscillators in the limit $N \to \infty$; the quantity $\rho(\theta, \omega)d\theta$ gives among those oscillators with frequency ω the fraction that have their phase between θ and $\theta + d\theta$. The oscillators with $|\omega_i| > Kr$ do not settle into a fixed point, but continue to rotate at a nonuniform rate (they are referred to as "drifting" oscillators). Concomitantly, on the unit circle, the corresponding phase points would be buzzing around the circle, spending naturally longer duration at locations that allow for a smaller local velocity $v(\theta, \omega)$ and zipping through locations that have a larger local velocity (Here, $v(\theta, \omega)$, the local velocity at position θ of oscillators that have natural frequency equal to ω, may be read off from Eq. (1.43) to be $v(\theta, \omega) = \omega - Kr \sin\theta$). Consequently, the density of this group of "drifting" oscillators would for most times be peaked around locations

---

[10]A fixed point exists also for the particular case $|\omega_i| = Kr$, for which $\theta(\omega_i) = \pi/2$ or $\theta(\omega_i) = 3\pi/2$ depending on the sign of ω_i. However, this fixed point is not stable. In the language of dynamical systems theory, $\omega_i = Kr$ that marks the boundary between two different behaviors corresponds to a saddle-node bifurcation. Since we are considering a continuum of frequencies distributed according to $g(\omega)$, the behavior for a single value of ω is not relevant for the final result.

with small local velocities, thus leading to a stationary density for this group that is inversely proportional to the local velocity:

$$\rho_{\mathrm{dr}}(\theta, \omega; r) \propto \frac{1}{|\omega - Kr \sin \theta|}. \tag{1.44}$$

The subscript reminds that this form is valid only for $|\omega| > Kr$. The constant of proportionality, which may be computed by imposing the normalization of ρ, is equal to $\sqrt{\omega^2 - (Kr)^2}/(2\pi)$. However, the factor of normalization is not necessary for our further steps, as will soon be clear. As remarked, the oscillators with $|\omega| < Kr$ will assume phase values equal to the fixed-point value $\theta(\omega)$ defined above. These oscillators are "locked"; in fact, recalling the argument presented in Sect. 1.6, we realize that the corresponding phases maintain a constant phase difference among them ("phase locking"), while evolving in time at a common frequency ("frequency locking") equal to 0 in the rotating frame and equal to ω_0 in the laboratory frame. The important point to note is that the overall distribution of the oscillators depends implicitly on the yet undetermined modulus r of the order parameter, and for this reason, we have added the parametric dependence on r in Eq. (1.44). The value of r has to be determined in a self-consistent way; the procedure goes as follows.

In the continuum description we are using, the relation (1.40) giving the order parameter becomes

$$re^{i\psi} = \int d\omega \int_0^{2\pi} d\theta \, g(\omega)e^{i\theta}\rho(\theta, \omega; r). \tag{1.45}$$

Let us first consider the contribution to this integral from frequencies with modulus larger than Kr; It is given by

$$\int_{|\omega| > Kr} d\omega \int_0^{2\pi} d\theta \, g(\omega)e^{i\theta}\rho_{\mathrm{dr}}(\theta, \omega; r). \tag{1.46}$$

The above expression vanishes due to the assumed symmetry of the frequency distribution, $g(\omega) = g(-\omega)$, and due to the relation $\rho_{\mathrm{dr}}(\theta + \pi, -\omega; r) = \rho_{\mathrm{dr}}(\theta, \omega; r)$, which can be easily deduced from Eq. (1.44). For the contribution of frequencies with modulus smaller than Kr, we use the fact that for each such oscillator, θ is fixed at a value that depends univocally on its frequency; we have

$$re^{i\psi} = \int_{|\omega| < Kr} d\omega \, g(\omega)e^{i\theta(\omega)}, \tag{1.47}$$

where $\theta(\omega) = \arcsin(\omega/Kr)$. With a change of integration variable, we get

$$re^{i\psi} = Kr \int_{-\frac{\pi}{2}}^{\frac{\pi}{2}} d\theta \, \cos\theta \, g(Kr \sin\theta)e^{i\theta}. \tag{1.48}$$

The imaginary part of the integral on the right hand side vanishes due to the symmetry of $g(\omega)$, thus giving $\psi = 0$. The real part is the final self-consistent relation for r

$$r = Kr \int_{-\frac{\pi}{2}}^{\frac{\pi}{2}} d\theta \, \cos^2\theta \, g(Kr\sin\theta). \tag{1.49}$$

The self-consistent equation (1.49) has a solution $r = 0$ that is always possible. From Eq. (1.44), we see that in this case, $\rho(\theta, \omega)$ is constant in θ for any ω; normalization requires that this constant is equal to $1/2\pi$. Another solution, namely, $r \neq 0$, to the self-consistent equation exists if the equation

$$1 = K \int_{-\frac{\pi}{2}}^{\frac{\pi}{2}} d\theta \, \cos^2\theta \, g(Kr\sin\theta) \tag{1.50}$$

is satisfied for $0 < r \leq 1$. It is not difficult to evaluate the range of K for which this solution exists, and to understand its behavior. First of all, by taking the limit of the right hand side of Eq. (1.50) for $r \to 0^+$, we obtain that the solution bifurcates continuously from the solution $r = 0$ at

$$K_c = \frac{2}{\pi g(0)}. \tag{1.51}$$

Furthermore, denoting with $A(K, r)$ the right hand side of Eq. (1.50), we obtain the following relations:

$$A(K, 1) = K \int_{-\frac{\pi}{2}}^{\frac{\pi}{2}} d\theta \, \cos^2\theta \, g(K\sin\theta) = \int_{-K}^{K} dx \left(1 - \frac{x^2}{K^2}\right)^{\frac{1}{2}} g(u) < 1, \tag{1.52}$$

$$A_r(K, r) = K^2 \int_{-\frac{\pi}{2}}^{\frac{\pi}{2}} d\theta \, \cos^2\theta \, \sin\theta \, g'(Kr\sin\theta) < 0, \tag{1.53}$$

$$A_K(K, r) = \int_{-\frac{\pi}{2}}^{\frac{\pi}{2}} d\theta \, \cos^2\theta \, [g(Kr\sin\theta) + \sin\theta \, g'(Kr\sin\theta)]$$

$$= \int_{-\frac{\pi}{2}}^{\frac{\pi}{2}} d\theta \, \sin^2\theta \, g(Kr\sin\theta) > 0. \tag{1.54}$$

In these equations, the prime denotes the derivative with respect to the argument, while the subscripts in A indicate partial derivatives. The first relation, $A(K, 1) < 1$, comes from the normalization of g; the same relation shows that $A(K, 1)$ tends to 1 as $K \to \infty$. The second relation, $A_r(K, r) < 0$, is a consequence of the fact that g is unimodal. The third relation, $A_K(K, r) > 0$, is evident. Taken together, the relations prove that Eq. (1.50) has a solution in r for $K \geq K_c$, that this solution is equal to 0 for $K = K_c$, that it increases with K, and that it approaches $r = 1$ as $K \to \infty$. For K close to K_c, one can obtain an explicit expression of r: By power expanding Eq. (1.50) for small r, we have

$$
\begin{aligned}
1 &= K \int_{-\frac{\pi}{2}}^{\frac{\pi}{2}} d\theta \left[g(0) \cos^2 \theta + \frac{1}{2} K^2 r^2 g''(0) \cos^2 \theta \sin^2 \theta \right] + O(r^4) \\
&= \frac{K}{K_c} + \frac{1}{16} \pi K^3 r^2 g''(0) + O(r^4),
\end{aligned}
\tag{1.55}
$$

where the expression (1.51) for K_c has been used in obtaining the second equality. By defining $\Delta K \equiv K - K_c$, one obtains at lowest order in ΔK the result

$$
r = \frac{4}{K_c^2} \sqrt{-\frac{\Delta K}{\pi g''(0)}},
\tag{1.56}
$$

(we note that for the class of functions $g(\omega)$ considered here, the second derivative at $\omega = 0$ is negative[11]). In the following chapter, treating the more general case with noise, we will show that the solution $r = 0$ is unstable for $K > K_c$, when the solution $r > 0$ is present.

1.7.2   The Ott-Antonsen Solution

In this section, we discuss an alternative analytical treatment of the Kuramoto model via the introduction and the implementation of the so-called Ott-Antonsen (OA) Ansatz [18, 19], which allows to rewrite in the thermodynamic limit the dynamics of coupled networks of phase oscillators in terms of a few collective variables. Specifically, in the context of the Kuramoto model (1.16) with a Lorentzian distribution of the oscillator frequencies, the ansatz studies the evolution in phase space by considering in the space $\mathscr{D}$ of all possible phase-space distributions $\rho(\theta, \omega, t)$ a particular class defined on and remaining confined to a manifold $\mathscr{M}$ in $\mathscr{D}$ under the time evolution of the phases. As a result of the choice of the particular class of $\rho(\theta, \omega, t)$, one obtains a single first-order ordinary differential equation for the evolution of the synchronization order parameter $r(t)$. The power and the usefulness of the ansatz lies in its remarkable ability to capture precisely and quantitatively through this single equation all, and not just some, of the order parameter attractors and bifurcations of the dynamics (1.16) (which may be obtained by performing numerical integration of the N coupled nonlinear equations (1.16) for $N \gg 1$ and evaluating $r(t)$ in numerics), for a Lorentzian $g(\omega)$. The success of the approach has

---

[11]Actually, it could be equal to 0. We will consider this case in the next Chapter when treating a uniform distribution.

led to hundreds of publications in applied mathematics and physics; a few recent ones are Refs. [20–23].

We begin the discussion by considering the case of a Lorentzian frequency distribution

$$g(\omega) = \frac{a}{\pi} \frac{1}{a^2 + \omega^2},$$ (1.57)

where a gives the width of the distribution. Substituting in Eq. (1.50), we have

$$1 = \frac{Ka}{\pi} \int_{-\frac{\pi}{2}}^{\frac{\pi}{2}} d\theta \, \cos^2 \theta \frac{1}{a^2 + K^2 r^2 \sin^2 \theta} = \frac{K}{a\pi} \int_{-1}^{1} dy \frac{\sqrt{1 - y^2}}{1 + b^2 y^2},$$ (1.58)

where in the second integral, we have made the change of variable $y = \sin \theta$, and have defined $b = Kr/a$. The last integral can be computed in closed form, using the indefinite integral expression

$$\int dy \frac{\sqrt{1 - y^2}}{1 + b^2 y^2} = \frac{\sqrt{1 + b^2}}{b^2} \arctan \frac{y\sqrt{1 + b^2}}{\sqrt{1 - y^2}} - \frac{1}{b^2} \arcsin y.$$ (1.59)

We thus obtain

$$1 = \frac{a}{Kr^2} \left(\sqrt{1 + \frac{K^2 r^2}{a^2}} - 1 \right),$$ (1.60)

which can be solved for r to give

$$r = \sqrt{1 - \frac{2a}{K}} = \sqrt{1 - \frac{K_c}{K}},$$ (1.61)

since Eq. (1.51) in this case gives $K_c = 2a$. We thus see that in this particular case, the order parameter can be obtained in closed form for any $K \geq K_c$.

However, the main purpose of this section is to show that under certain assumptions and for a Lorentzian frequency distribution, the full time-dependent dynamics of $r(t)$ can be determined. The procedure, introduced by Ott and Antonsen, goes as follows: Using from the start the continuum description of the Kuramoto dynamics via the distribution function $\rho(\theta, \omega, t)$, the equations of motion (1.16) corresponds to a continuity equation that describes the conservation under the dynamical evolution of the total number of oscillators with a given frequency:

$$\frac{\partial \rho(\theta, \omega, t)}{\partial t} = -\frac{\partial}{\partial \theta} \left[(\omega + F(\theta, t)) \, \rho(\theta, \omega, t) \right],$$ (1.62)

with

$$F(\theta, t) = K \int d\omega \int_0^{2\pi} d\theta' \, g(\omega) \sin(\theta' - \theta) \rho(\theta', \omega, t).$$ (1.63)

Using the complex order parameter $\tilde{r} = re^{i\psi}$ given in the continuum limit by

$$\tilde{r} = \int d\omega \int_0^{2\pi} d\theta\, g(\omega)e^{i\theta}\rho(\theta, \omega, t), \tag{1.64}$$

(where the time dependence of $\tilde{r}$ is not explicitly indicated), the continuity equation may be written as

$$\frac{\partial \rho(\theta, \omega, t)}{\partial t} = -\frac{\partial}{\partial \theta}\left\{\left[\omega + \frac{K}{2i}\left(\tilde{r}e^{-i\theta} - \tilde{r}^*e^{i\theta}\right)\right]\rho(\theta, \omega, t)\right\}, \tag{1.65}$$

where the star denotes complex conjugation.

Since $\rho(\theta, \omega, t)$ is 2π-periodic in θ for all ω and t, we may perform a Fourier series expansion of $\rho(\theta, \omega, t)$:

$$\rho(\theta, \omega, t) = \frac{1}{2\pi}\sum_{n=-\infty}^{+\infty}\tilde{\rho}_n(\omega, t)e^{in\theta}, \tag{1.66}$$

with $\tilde{\rho}_{-n}(\omega, t) = \tilde{\rho}_n^*(\omega, t)$ following from the fact that $\rho(\theta, \omega, t)$ is real, and with $\tilde{\rho}_0(\omega, t) \equiv 1$ because of the normalization of ρ. Substituting in Eq. (1.65), we get a system of equations

$$\frac{\partial \tilde{\rho}_n(\omega, t)}{\partial t} + in\omega\tilde{\rho}_n(\omega, t) + \frac{K}{2}n\left(\tilde{r}\tilde{\rho}_{n+1}(\omega, t) - \tilde{r}^*\tilde{\rho}_{n-1}(\omega, t)\right) = 0, \tag{1.67}$$

with

$$\tilde{r} = \int d\omega\, g(\omega)\tilde{\rho}_1^*(\omega, t). \tag{1.68}$$

The brilliant step introduced by Ott and Antonsen is to study the system of equation under the ansatz

$$\tilde{\rho}_n(\omega, t) = \alpha^n(\omega, t); \quad n \geq 0. \tag{1.69}$$

One has to assume also that $\alpha(\omega, t) \leq 1$, in order to have convergence of the Fourier series, Eq. (1.66). The system of equation then reduces to the single equation

$$\frac{\partial \alpha(\omega, t)}{\partial t} + i\omega\alpha(\omega, t) + \frac{K}{2}\left(\tilde{r}\alpha^2(\omega, t) - \tilde{r}^*\right) = 0, \tag{1.70}$$

with

$$\tilde{r} = \int d\omega\, g(\omega)\alpha^*(\omega, t). \tag{1.71}$$

Obviously the functions satisfying Eq. (1.69) are a restricted class, but Eq. (1.70) shows that if at time $t = 0$, a function belongs to this class, the subsequent evolution governed by the continuity equation continues to have the function in the same class.

It is also important that the property $\alpha(\omega, t) \leq 1$ is preserved by the dynamics.[12] The physical justification to study this class is that the stationary states studied in the previous section belong to it. In fact, by performing the summation on the right hand side of Eq. (1.66) by using Eq. (1.69), we get using the known expression for the summation of geometric series the result

$$\rho(\theta, \omega, t) = \frac{1}{2\pi} \frac{1 - |\alpha|^2}{(1 - |\alpha|)^2 + 4|\alpha| \sin^2\left(\frac{\theta - \phi}{2}\right)}, \qquad (1.72)$$

where $\alpha(\omega, t) = |\alpha| e^{-i\phi}$ (with the dependence on ω and t not explicitly shown in the modulus and the phase).

It is not difficult to see that the drifting solution of Eq. (1.44) is obtained for $\phi = \pi/2$ and $|\alpha|^2 = \left[2 - d^2 - 2\sqrt{1 - d^2}\right]/d^2$, where $d = (Kr)/\omega$. In particular, we have $|\alpha| = 0$ for $r = 0$, and the uniform distribution $1/(2\pi)$ is recovered. On the other hand, for $|\alpha| \to 1$, the distribution (1.72) tends to a δ-function in $\theta = \phi$, which describes the locked oscillators.

To proceed, Ott and Antonsen make further assumptions. The first is that the initial condition $\alpha(\omega, 0)$ can be analytically continued to the whole of the complex-ω plane, and that this continuation has no singularity in the lower-half complex-ω plane, with $\alpha(\omega, 0)$ tending to 0 when the imaginary part of ω tends to $-\infty$. The second is that the distribution $g(\omega)$ has only a finite (actually only a small) number of poles in this lower-half plane when considered as a function of complex ω. They prove that the dynamics (1.70) conserves these properties for $\alpha(\omega, t)$. Then, the theorem of residues applied to the integral in Eq. (1.71) allows to express $\tilde{r}$ in terms of the values of $\alpha(\omega, t)$ at the poles of $g(\omega)$ in the lower-half complex-ω plane. In the case of a Lorentzian, Eq. (1.57), $g(\omega)$ has only one pole in the lower-half plane, and we easily obtain

$$\tilde{r}(t) = \alpha^*(-ia, t). \qquad (1.73)$$

Writing Eq. (1.70) for $\omega = -ia$, taking the complex conjugate, and using $\tilde{r} = re^{i\psi}$, the imaginary part gives $d\psi/dt = 0$, while the real part gives

$$\frac{dr(t)}{dt} + ar(t) + \frac{K}{2}\left(r^3(t) - r(t)\right) = 0. \qquad (1.74)$$

We see that the assumptions that have been introduced allow to obtain a single ordinary differential equation for the order parameter. Equation (1.74) may be integrated to get

---

[12]The reader is referred to the original work by Ott and Antonsen, Refs. [18, 19], for the proof of this property.

$$r^2(t) = \frac{(K - 2a)\, r^2(0)}{\left(K - 2a - Kr^2(0)\right) e^{(2a-K)t} + Kr^2(0)}; \quad K \neq 2a \qquad (1.75)$$

$$r^2(t) = \frac{r^2(0)}{1 + 2atr^2(0)}; \quad K = 2a. \qquad (1.76)$$

These equations show that for $K > K_c = 2a$, the order parameter $r(t)$ tends exponentially fast to the stationary-state value found in the previous section, Eq. (1.61), while for $K < K_c$, it tends exponentially fast to 0 (a power-law relaxation occurs for $K = K_c$).

1.8 The Oscillators with Inertia

Until now we have considered a first-order dynamics for the interacting oscillators, of the type (1.30). As we have explained, the interacting oscillators of synchronizing systems are complex dynamical systems that have the dynamics of a limit cycle, and the effective description with first-order dynamics is the natural one in this case.

Since we are using an effective description of complex systems, thus introducing a series of approximations, it is possible that some features are lost, especially in the process that leads the system from an initial state to a synchronized state. In particular, it has been observed that the approach to synchronization is often slower than that predicted by the simplified models. A possible route to obtain a slower approach to the synchronized state is to elevate the order of the dynamics, from equations of motion that are first order in time to those that are second order in time. Before treating the specific case of the oscillators of synchronizing systems, let us compare in general a first-order and a second-order dynamics by considering the damped motion of a particle of mass m moving on a circle of unit length, and subject to a torque τ and a restoring force to $\theta = 0$. The equation of motion is

$$m\frac{\mathrm{d}^2\theta}{\mathrm{d}t^2} = -\gamma\frac{\mathrm{d}\theta}{\mathrm{d}t} + \gamma\omega - \gamma\sin\theta, \qquad (1.77)$$

where γ is the friction coefficient. The reason why we have chosen the intensity of the restoring force to be numerically equal to the friction coefficient will be clear in a moment. For the same reason, we also pose $\tau \equiv \gamma\omega$. If the mass m is very small, the acceleration of θ is very large, unless the right hand side of Eq. (1.77) is also very small. Therefore, we expect that starting from a generic initial condition specified by the initial values of θ and $\dot{\theta} \equiv \mathrm{d}\theta/\mathrm{d}t$, the motion will soon approach a region of the dynamical phase space where the right hand side of the equation is small. In the limiting case of $m = 0$, the equation of motion becomes first-order in time:

$$\frac{\mathrm{d}\theta}{\mathrm{d}t} = \omega - \sin\theta. \qquad (1.78)$$

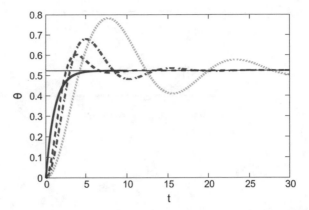

Fig. 1.4 The dynamics $\theta(t)$ determined by Eq. (1.77) with $\omega = 0.5$ and γ/m equal to 0.2 (dotted line), 0.5 (dot-dashed line), and 1.0 (dashed line). The initial condition is $\theta(0) = 0$ and $\dot{\theta}(0) = 0$. The full curve is the dynamics determined by Eq. (1.78) with $\omega = 0.5$ and initial condition $\theta(0) = 0$. The horizontal full line shows the asymptotic fixed value equal to $\pi/6$

To see what happens with the introduction of mass, we now compare the numerical solution of Eq. (1.78) with that of Eq. (1.77) for three different values of γ/m, i.e., $\gamma = 0.2, 0.5$ and 1.0; the larger this value, the smaller is the mass. For definiteness, we choose $\omega = 0.5$. We have seen before that in this case, the dynamics determined by Eq. (1.78) tends to a fixed point equal to $\pi/6$ (this is why we have put a coefficient equal to γ in the restoring force in Eq. (1.77), which allows to compare with the case already studied). In Fig. 1.4, we plot the dynamics determined by Eq. (1.77) for the three mentioned values of γ/m and for the initial condition given by $\theta(0) = 0$ and $\dot{\theta}(0) = 0$, and the dynamics (1.78) determined by Eq. (1.78) for the initial condition $\theta(0) = 0$. The figure shows that in all cases, the dynamics approaches the asymptotic fixed-point value equal to $\pi/6$, with the overdamped ($m = 0$) dynamics being faster in reaching the fixed value, and with the time needed to approach this value increasing with the increase of mass. Physically, a larger mass, i.e., a larger inertia, causes a slower approach towards the fixed point.

We may hope this aforementioned slower response occurs also with respect to the approach to synchronization. Let us then consider the case of two oscillators in interaction with the dynamics given by Eq. (1.37). We compute the numerical solution of those equations together with that of the corresponding dynamics when the oscillators have an inertia. Thus, we consider the equations of motion

$$m\frac{d^2\theta_1}{dt^2} = -\gamma\frac{d\theta_1}{dt} + \gamma\omega_1 - \gamma\sin(\theta_1 - \theta_2), \quad m\frac{d^2\theta_2}{dt^2} = -\gamma\frac{d\theta_2}{dt} + \gamma\omega_2 + \gamma\sin(\theta_1 - \theta_2). \quad (1.79)$$

For simplicity, we have assumed the same mass for the two oscillators. In this example, we take the values $\omega_1 = 1$ and $\omega_2 = 1.1$. In Fig. 1.5, we plot $\phi(t) \equiv \theta_1(t) - \theta_2(t)$ determined by these equations for the three cases with γ/m equal to 0.2, 0.5 and 2.0, together with $\phi(t)$ as determined by Eq. (1.37).

In all cases, $\phi(t)$ tends to the asymptotic value $\phi^* = \arcsin\left[(\omega_1 - \omega_2)/2\right]$ $= -\arcsin(0.05)$; the phase difference between the two oscillators locks at this value, and the oscillators get synchronized, rotating with the same frequency

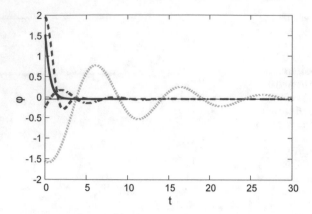

Fig. 1.5 The time evolution of $\phi(t) = \theta_1(t) - \theta_2(t)$ determined by Eq. (1.79) with $\omega_1 = 1.0$ and $\omega_2 = 1.1$, with γ/m equal to 0.2 (dotted line), 0.5 (dot-dashed line), and 2.0 (dashed line). The initial condition $\theta(0)$ and $\dot{\theta}(0)$ have been chosen randomly in each case. The full curve is $\phi(t)$ determined by Eq. (1.37), again with random initial condition $\theta(0)$. The horizontal full line shows the asymptotic fixed value equal to $-\arcsin(0.05)$

$\omega_0 = (\omega_1 + \omega_2)/2$. The plot shows that the synchronization is faster for the case without inertia ($m = 0$), while it becomes slower and slower by increasing the mass.

It is then plausible that even in systems with many oscillators, the introduction of an inertia can slow down the approach to synchronization. It is important to keep in mind that the introduction of inertia has to be considered only as an operative approach that is useful in the study of synchronizing systems. However, we have emphasized that each oscillator of a synchronizing system is actually a complex system that is represented by a single dynamical degree of freedom under an effective description that by its very nature is approximate. Therefore, it is operationally justified to implement a simple procedure, e.g., introducing inertia, if this has as a consequence a better agreement of the model with the dynamics of the approach to synchronization.

Although the procedure is conceptually simpler, it can have important effects on the nature of the transitions from an unsynchronized to a synchronized state, changing the transition from second order to first. This will be studied in details in Chap. 3 devoted to systems with second-order dynamics. As for systems with first-order dynamics, we will consider synchronizing systems with noise that we will study with tools of statistical physics.

Appendix 1: A Two-Dimensional Dynamics with a Limit-Cycle Attractor

Let us consider the following example of a two-dimensional dynamics that represent the so-called Stuart-Landau model:

$$\frac{dx_1}{dt} = \mu x_1 - \omega x_2 - \beta(x_1^2 + x_2^2)x_1, \quad \frac{dx_2}{dt} = \omega x_1 + \mu x_2 - \beta(x_1^2 + x_2^2)x_2, \quad (1.80)$$

where ω and β are fixed parameters, while μ represents a parameter that can be tuned. It is evident that $(0, 0)$ is a fixed point of the dynamics (1.80). Its stability is studied by linearizing Eq. (1.80) about the fixed point. The linearized equations read

$$\frac{dx_1}{dt} = \mu x_1 - \omega x_2, \quad \frac{dx_2}{dt} = \omega x_1 + \mu x_2. \quad (1.81)$$

The eigenvalues of the associated linear time-evolution operator are $\lambda = \mu \pm i\omega$. We conclude that the fixed point $(0, 0)$ is stable for $\mu < 0$ and is unstable for $\mu > 0$. To study the case $\mu > 0$, we may choose to go to the polar coordinates (r, ϕ), and rewrite the equations as

$$\frac{dr}{dt} = \mu r - \beta r^3, \quad \frac{d\phi}{dt} = \omega. \quad (1.82)$$

From the above equations, it is evident that ϕ is a neutrally-stable variable, while the dynamics of r has two fixed points: $r = 0$, which is unstable, and $r = \sqrt{\mu/\beta}$, which is stable. We thus have a limit cycle, which in the present situation is a circle of radius equal to $\sqrt{\mu/\beta}$. We conclude that this circle is an attractor: all orbits (excepting the one staying exactly on the unstable origin) will be attracted to this circle by the dynamics. Readers acquainted with dynamical systems theory have surely recognized in this example the standard description of a Hopf bifurcation [7], which occurs in the passage from $\mu < 0$ to $\mu > 0$, and which marks the change from a situation with a stable fixed point to one in which the fixed point becomes unstable and a stable limit cycle appears.

Appendix 2: The Lyapunov Exponents

The Lyapunov exponents of a dynamical system characterize the rate of separation in time of initially infinitesimally-close trajectories in the phase space of the system [24]. For an n-dimensional dynamical system $dx_i/dt = F_i(x_1, x_2, \ldots, x_n)$, one has the spectrum of Lyapunov exponents $\lambda_1, \lambda_2, \ldots, \lambda_n$. For such a system, two trajectories in phase space with an initial separation equal to $d(0)$ would separate out in time, with the separation at long times behaving typically as

$$d(t) \approx e^{\lambda t} d(0), \tag{1.83}$$

where λ is the so-called maximal Lyapunov exponent, defined as the maximum of the n Lyapunov exponents. In general, λ (as also the whole spectrum) depends on the initial phase-space points of the trajectories. A positive λ implies chaoticity. The trajectories close and tending to the limit cycle have all but one negative Lyapunov exponents, while the maximum exponent is zero. These trajectories are thus not chaotic. As for any dynamics, the zero exponent is associated with perturbations along the trajectory, which for a limit cycle corresponds to a perturbation of the phase.

Appendix 3: The One-Body Distribution Function in an N-Body System

In this appendix, we clarify the issue of the interpretation of the one-body distribution function that is generally employed when a comparison is made with the results of a numerical simulation or with the results of an experimental measurement. The issue is common to the study of any system with a large number of components irrespective of the properties of interaction between the components. Therefore, we choose to consider the case of an isolated Hamiltonian system, and for simplicity, we consider the particles to be moving in one dimension to avoid vector notations (our argument is independent of the embedding dimension).

We assume to have a system of N particles moving in one dimension that are described by the canonical coordinates $\{q_i, p_i\}$, where q_i is the Cartesian coordinate of the i-th particle and p_i is its momentum. In the statistical approach, we have an N-body distribution function $\rho_N(q_1, \ldots, q_N, p_1, \ldots, p_N, t)$ representing an ensemble of such systems. The distribution function evolves according to the Liouville equation

$$\frac{\partial \rho_N}{\partial t} + \sum_{i=1}^{N} \left[\frac{\partial \rho_N}{\partial q_i} \frac{\partial H}{\partial p_i} - \frac{\partial \rho_N}{\partial p_i} \frac{\partial H}{\partial q_i} \right] = 0, \tag{1.84}$$

where $H(q_1, \ldots, q_N, p_1, \ldots, p_N)$ is the Hamiltonian of the system. Most important of all are the so-called reduced distribution functions, the simplest of which being the one-body distribution function $\rho(q, p, t)$, defined by

$$\rho(q, p, t) = \int dq_2 \ldots dq_N dp_2 \ldots dp_N \, \rho(q, q_2, \ldots, q_N, p, p_2, \ldots, p_N, t). \tag{1.85}$$

This definition exploits the invariance of the Hamiltonian with respect to particle permutations that is generally assumed. It turns out that for a generic system, the time-evolution for an s-body reduced distribution function depends explicitly on the $(s + 1)$-body distribution function, thus forming a coupled chain of infinite number of equations that goes under the name of the so-called Bogoliubov-Born-Green-Kirkwood-Yvon (BBGKY) hierarchy [25]. Here, we are not interested in the equation that describes the time evolution of $\rho(q, p, t)$, since there are several such equations resulting from different approximations of closing the hierarchy that are suitable for different classes of systems. Here, instead, we are interested in the interpretation of the function $\rho(q, p, t)$.

According to the meaning of an ensemble of systems, it follows that choosing a given particle of the system, e.g., the one labeled by i, the function $\rho(q, p, t)$ gives the distribution at time t, over the members of the ensemble, of the position and momentum of the i-th particle. The fact that this distribution is the same for any i is a consequence of the permutation invariance of the Hamiltonian. As a result, if $h(q, p)$ is a function of the position and momentum of a particle, its ensemble average is given by

$$\langle h \rangle (t) = \int dq dp \, \rho(q, p, t) h(q, p). \tag{1.86}$$

However, if one obtains $\rho(q, p, t)$ from its evolution equation, and at the same time performs a numerical experiment in which one realization of the dynamics of the system is simulated, the comparison that is generally done is between $\langle h \rangle (t)$ and

$$h_m(t) = \frac{1}{N} \sum_{i=1}^{N} h(q_i(t), p_i(t)), \tag{1.87}$$

where the subscript in $h_m(t)$ denotes the average over the N particles of the system. The quantities $\langle h \rangle (t)$ and $h_m(t)$ are in principle different quantities: the former is the ensemble average of the function $h(q, p)$ computed for a given particle, while the latter is the arithmetic average over all the particles of only one member of the ensemble of systems. Nevertheless, they are usually compared, for example when a numerical simulation is performed and the accuracy of a theoretical evaluation is investigated.[13] A similar thing occurs in a comparison between theory and an experimental measurement, where operatively what is measured is $h_m(t)$. This identification is done also in the very important case of a stationary distribution $\rho(q, p)$, where a theoretical evaluation is considered satisfactory if the time-dependent quantity $h_m(t)$ has actually only very small fluctuations around a value, the latter in turn being very close to the time-independent quantity $\langle h \rangle$.

As a very simple example, we can consider the evaluation related to the quantity $h = 1/(2M) \left(p_x^2 + p_y^2 + p_z^2 \right)$, where M is the mass of a particle (for a concrete example, we go back to a three-dimensional system). This is of course related to the determination of the average kinetic energy of a particle, which in turn is related to the temperature of the system. Assume for example an almost ideal gas of N

---

[13]Sometimes the quantity $h_m(t)$ is obtained by further averaging over numerical simulation of several realizations of the dynamics, thus going in the direction of performing an ensemble average. This does not affect our argument, since this further average is employed to decrease the statistical fluctuations, but the main averaging procedure is the one over the $N \gg 1$ particles of the system.

particles, so that the potential energy is very small compared to the kinetic energy. If we consider an isolated system at energy E, then we expect that in the stationary state, we have $\langle h \rangle = E/N$. Running a simulation and evaluating $h_m(t)$, we will find that, possibly after a transient, this quantity will remain close to E/N, with very small fluctuations around this value.

The property that $h_m(t)$ and $\langle h \rangle(t)$ are very close, actually coinciding with probability one when N goes to infinite, is called self-averaging. There are situations in which it does not hold, especially in disordered systems. Further comments on this issue are outside the scope of this monograph.

References

1. S.H. Strogatz, *Sync: The Emerging Science of Spontaneous Order* (Hyperion, New York, 2003)
2. A. Pikovsky, M. Rosenblum, J. Kurths, *Synchronization: A Universal Concept in Nonlinear Sciences* (Cambridge University Press, Cambridge, 2001)
3. S.H. Strogatz, I. Stewart, Sci. Am. **269**, 102 (1993)
4. M. Rosenblum, A. Pikovsky, Contem. Phys. **44**, 401 (2003)
5. A. Pikovsky, M. Rosenblum, Scholarpedia **2**(12), 1459 (2007)
6. R. Livi, P. Politi, *Nonequilibrium Statistical Physics: A Modern Perspective* (Cambridge University Press, Cambridge, 2017)
7. S.H. Strogatz, *Nonlinear Dynamics and Chaos: With Applications to Physics, Biology, Chemistry, and Engineering* (Westview Press, Boulder, 2014)
8. E. Ott, *Chaos in Dynamical Systems* (Cambridge University Press, Cambridge, 2002)
9. A. Campa, T. Dauxois, D. Fanelli, S. Ruffo, *Physics of Long-Range Interacting Systems* (Oxford University Press, Oxford, 2014)
10. Y. Kuramoto, in International Symposium on Mathematical Problems in Theoretical Physics, ed. by H. Araki. Lecture Notes in Physics, vol. 39 (Springer, New York, 1975)
11. Y. Kuramoto, *Chemical Oscillations, Waves and Turbulence* (Springer, Berlin, 1984)
12. S.H. Strogatz, Physica D **143**, 1 (2000)
13. J.A. Acebrón, L.L. Bonilla, C.J.P. Vicente, F. Ritort, R. Spigler, Rev. Mod. Phys. **77**, 137 (2005)
14. S. Gupta, A. Campa, S. Ruffo, J. Stat. Mech. Theory Exp. R08001 (2014)
15. A. Pikovsky, M. Rosenblum, Chaos **25**, 097616 (2015)
16. H. Risken, *The Fokker-Planck Equation: methods of Solution and Applications* (Springer, Berlin, 1996)
17. H. Sakaguchi, Prog. Theor. Phys. **79**, 39 (1988)
18. E. Ott, T.M. Antonsen, Chaos **18**, 037113 (2008)
19. E. Ott, T.M. Antonsen, Chaos **19**, 023117 (2009)
20. O.E. Omel'chenko, M. Wolfrum, C.R. Laing, Chaos **24**, 023102 (2014)
21. D. Pazó, E. Montbrió, Phys. Rev. Lett. **116**, 238101 (2016)
22. K.P. O'Keeffe, S.H. Strogatz, Phys. Rev. E **93**, 062203 (2016)
23. X. Zhang, A. Pikovsky, Z. Liu, Sci. Rep. **7**, 2104 (2017)
24. A. Pikovsky, A. Politi, *Lyapunov Exponents: a Tool to Explore Complex Dynamics* (Cambridge University Press, Cambridge, 2016)
25. K. Huang, *Statistical Mechanics* (Wiley, New York, 1987)

Chapter 2
Oscillators with First-Order Dynamics

Abstract In the first section, we derive the Fokker-Planck equation that determines the dynamics of the one-body distribution function associated with a system of interacting oscillators with distributed natural frequencies. In the second section, we discuss the celebrated Kuramoto model; most results have been obtained for this model, which constitutes the basic model studied in this monograph. Introduced originally in the noiseless case, we study it in the presence of noise. In the third section, we study the Kuramoto model in the case where the frequency distribution is symmetric and unimodal. In the fourth section, we take up the case of a bimodal yet symmetric frequency distribution, and show that there is a much richer set of possible long-time states of the dynamics. In the fifth and final section, we describe the results that have been obtained in models that extend and generalize the Kuramoto model, by considering generic interactions between the oscillators.

Keywords Oscillators · First-order dynamics · Fokker-Planck equation
Order parameter · Unimodal frequency distribution · Critical coupling constant
Bimodal frequency distribution

In this chapter, we begin with our analysis of a collection of interacting oscillators subject to noise. As we have seen in Chap. 1, each oscillator may be effectively described with a phase variable, not only when it is isolated, but also when it is in interaction with other oscillators, provided that the interaction is weak enough so that its effect may be reduced to a modification of the phase dynamics. In addition, we have seen that the interaction between each pair of oscillators may be accounted for by a function of the difference of their phases. Assuming furthermore that the interaction is the same for each pair of oscillators, we have thus arrived at the equations of motion (1.15).

Our purpose in this chapter is to consider such coupled equations in the presence of noise. In Eq. (1.30), we have written the corresponding coupled Langevin equations in the case in which all the natural frequencies are equal; now we study the general case in which the natural frequencies are distributed according to a given distribution function. Our first task is to obtain the Fokker-Planck equation for the one-body distribution function, which, as emphasized earlier, is the main tool for our statistical physics description of synchronizing systems.

2.1   The Oscillators with Distributed Natural Frequencies

Let us start with the coupled Langevin equations for the dynamics of noisy interacting oscillators given by

$$\frac{d\theta_i}{dt} = \omega_i + \frac{K}{N} \sum_{j=1}^{N}{}' f(\theta_i - \theta_j) + \eta_i(t), \quad i = 1, 2, \ldots, N, \tag{2.1}$$

where $\eta_i(t)$ is the stochastic noise; as noted at the beginning of Sect. 1.5, it may be regarded to account for the stochastic fluctuations of the frequencies in time. Here, the prime in the sum indicates the fact that the term with $i = j$ is to be excluded while performing the sum. The natural frequencies ω_i's for the different oscillators are extracted independently from a given distribution function $g(\omega)$ with the normalization $\int_{-\infty}^{\infty} d\omega \, g(\omega) = 1$. The noise $\eta_i(t)$'s for the different oscillators are taken to be independent, and we consider them to be Gaussian distributed at a given time and uncorrelated between different times. Thus, the noise has the following expectation values:

$$\langle \eta_i(t) \rangle = 0, \quad \langle \eta_i(t)\eta_j(t') \rangle = 2D\delta_{ij}\delta(t - t'). \tag{2.2}$$

2.1.1   Derivation of the Fokker-Planck Equation

A description of a given Langevin dynamics via a Fokker-Planck equation involves considering at a conceptual level an average of the dynamics over its many realizations, i.e., over many realizations of the noise [1]. This description is obtained with the introduction of a time-dependent distribution function of the phases of the oscillators, and is most appropriate in the case when the number N of oscillators is extremely large, ideally as $N \to \infty$.

For a finite but very large number of oscillators, let us consider the distribution function $\rho_N(\theta_1, \theta_2, \ldots, \theta_N, t)$, defined as the probability density at time t to have the phase of the first oscillator equal to θ_1, the phase of the second oscillator equal to θ_2, the phase of the third oscillator equal to θ_3, and so on. The natural frequency of each oscillator is extracted from the distribution function $g(\omega)$, and by relabeling the oscillators, if necessary, we may take the first n_1 of them to have their natural frequency equal to ω_1, the next n_2 of them to have their natural frequency equal to ω_2, and so on, up to the last n_p number of oscillators that have their natural frequency equal to ω_p. Of course, we have[1] $n_1 + n_2 + \cdots + n_p = N$. We take the function ρ_N to be normalized, i.e.,

---

[1] With a finite yet large number of oscillators, we will obviously have a discrete set of frequency values extracted from $g(\omega)$.

$$\int_0^{2\pi} d\theta_1 \int_0^{2\pi} d\theta_2 \ldots \int_0^{2\pi} d\theta_N \; \rho_N(\theta_1, \theta_2, \ldots, \theta_N) = 1, \tag{2.3}$$

and to be symmetric with respect to permutations of the θ's within each group with a given natural frequency. The evolution of the function ρ_N is given by the N-body Fokker-Planck equation that may be written down straightforwardly from Eq. (2.1) using standard procedure as [1]

$$\frac{\partial \rho_N}{\partial t} = -\sum_{i=1}^{N} \omega_i \frac{\partial \rho_N}{\partial \theta_i} - \frac{K}{2N} \sum_{i=1}^{N} \sum_{j=1}^{N} \left(\frac{\partial}{\partial \theta_i} - \frac{\partial}{\partial \theta_j} \right) \left[f(\theta_i - \theta_j) \rho_N \right] + D \sum_{i=1}^{N} \frac{\partial^2 \rho_N}{\partial \theta_i^2}. \tag{2.4}$$

From Eq. (2.4), we want to obtain an equation for the one-body distribution function that characterizes the probability density to obtain any of the θ's within a given group to assume a given value at a given time. To this end, we introduce the so-called reduced distribution functions [2], which are obtained from ρ_N by integrating out a subset of the θ's, thereby yielding a function of the remaining θ's. For example, for $k = 1, 2, \ldots, p$, by integrating out $(n_k - s_k)$ number of θ variables belonging to the group with natural frequency ω_k, we get a reduced distribution function that depends on the s_k number of θ variables in the group. With $N_k \equiv \sum_{r=1}^{k} n_r$, the formal expression for the reduced distribution function reads

$$\rho_{s_1, s_2, \ldots, s_p}(\theta_1, \ldots, \theta_{s_1}, \theta_{N_1+1}, \ldots, \theta_{N_1+s_2}, \theta_{N_2+1}, \ldots, \theta_{N_{p-1}+s_p}, t)$$

$$= \left[\prod_{r=1}^{p} \left(\frac{n_r!}{(n_r - s_r)! n_r^{s_r}} \right) \right]$$

$$\times \int d\theta_{s_1+1} \ldots d\theta_{N_1} d\theta_{N_1+s_2+1} \ldots d\theta_{N_2} \ldots d\theta_{N_{p-1}+s_p+1} \ldots d\theta_N \; \rho_N(\theta_1, \ldots, \theta_N, t). \tag{2.5}$$

Here, we have adopted an often used factor of proportionality in the second line of the expression that guarantees the normalization of the reduced distribution function. The symmetry of ρ_N with respect to permutations of the θ's within each of the p groups makes it irrelevant with respect to which of the $(n_k - s_k)$ variables of the k-th group, $k = 1, 2, \ldots, p$, do we perform the integration. The evolution equation of the reduced functions are obtained by integrating suitably Eq. (2.4).

In the following, we are interested in the evolution equation of one-body reduced distribution functions. For example, considering a variable of the first group, we are interested in the evolution of

$$\rho_{1,0,\ldots,0}(\theta_1, t) = \int d\theta_2 \ldots d\theta_N \; \rho_N(\theta_1, \ldots, \theta_N, t). \tag{2.6}$$

Note that this function and other one-body reduced functions are all normalized to unity. In the following computation, we will also require the two-body reduced distribution functions, which may describe the probability density for obtaining two θ's of the same group or two θ's belonging to two different groups to assume given values at a given time. For example, we have

$$\rho_{2,0,\dots,0}(\theta_1, \theta_2, t) = \frac{n_1 - 1}{n_1} \int d\theta_3 \dots d\theta_N \, \rho_N(\theta_1, \dots, \theta_N, t), \qquad (2.7)$$

$$\rho_{1,1,0,\dots,0}(\theta_1, \theta_{n_1+1}, t) = \int d\theta_2 \dots d\theta_{n_1} d\theta_{n_1+2} \dots d\theta_N \, \rho_N(\theta_1, \dots, \theta_N, t). \quad (2.8)$$

Let us now derive the evolution equation for the one-body reduced functions. By integrating the Fokker-Planck equation (2.4) with respect to $\theta_2, \dots, \theta_N$, we get the equation for $\rho_{1,0,\dots,0}(\theta_1, t)$. The integration of the left hand side and of the first and the last term on the right hand side of Eq. (2.4) are trivial, and we get

$$\frac{\partial}{\partial t} \rho_{1,\dots,0}(\theta_1, t) = -\omega_1 \frac{\partial}{\partial \theta_1} \rho_{1,\dots,0}(\theta_1, t) + \text{interaction terms} + D \frac{\partial^2}{\partial \theta_1^2} \rho_{1,\dots,0}(\theta_1, t).$$
$$(2.9)$$

By integrating the middle term on the right hand side of Eq. (2.4) (in Eq. (2.9), we have referred to the result of integration as "interaction terms"), the only terms of the sum that give a non-vanishing contribution are those in which either i or j is equal to 1. We thus obtain from the integration of the middle term on the right hand side of Eq. (2.4) the following expression:

$$-\frac{K}{N} \sum_{j=2}^{N} \int d\theta_2 \dots d\theta_N \frac{\partial}{\partial \theta_1} \left[f(\theta_1 - \theta_j) \rho_N(\theta_1, \dots, \theta_N, t) \right]. \qquad (2.10)$$

By exploiting the symmetry of ρ_N with respect to permutations of the θ's within the same group, we may express the last expression as

$$-\frac{K}{N}(n_1 - 1) \int d\theta_2 \dots d\theta_N \frac{\partial}{\partial \theta_1} \left[f(\theta_1 - \theta_2) \rho_N(\theta_1, \dots, \theta_N, t) \right]$$
$$-\frac{K}{N} \sum_{r=1}^{p-1} n_{r+1} \int d\theta_2 \dots d\theta_N \frac{\partial}{\partial \theta_1} \left[f(\theta_1 - \theta_{N_r+1}) \rho_N(\theta_1, \dots, \theta_N, t) \right]. \quad (2.11)$$

By invoking the definition of the reduced distribution function, we rewrite the last expression as

$$-\frac{K}{N} n_1 \int d\theta_2 \frac{\partial}{\partial \theta_1} \left[f(\theta_1 - \theta_2) \rho_{2,0,\dots,0}(\theta_1, \theta_2, t) \right]$$
$$-\frac{K}{N} \sum_{r=1}^{p-1} n_{r+1} \int d\theta_{N_r+1} \frac{\partial}{\partial \theta_1} \left[f(\theta_1 - \theta_{N_r+1}) \rho_{1,0,\dots,0,1,0,\dots,0}(\theta_1, \theta_{N_r+1}, t) \right], \quad (2.12)$$

where in the last line the second "1" in the subscript is in the $(r + 1)$-th position.

We may now use the approximation of neglecting two-body correlations to write

$$\rho_{2,0,...,0}(\theta_1, \theta_2, t) \approx \rho_{1,0,...,0}(\theta_1, t)\rho_{1,0,...,0}(\theta_2, t), \tag{2.13}$$

$$\rho_{1,0,...,0,1,0,...,0}(\theta_1, \theta_{N_r+1}, t) \approx \rho_{1,0,...,0}(\theta_1, t)\rho_{0,...,0,1,0,...,0}(\theta_{N_r+1}, t). \tag{2.14}$$

This approximation is expected to be exact in the limit $N \to \infty$, and is a rather good approximation for large N. Moreover, this approximation is the simplest one may invoke to close the infinite hierarchy (the so-called Bogoliubov-Born-Green-Kirkwood-Yvon (BBGKY) hierarchy) arising from the fact that the equation for an s-particle reduced distribution function includes the $(s + 1)$-particle distribution function, thus forming a coupled chain of infinite number of equations [2]. Substituting Eqs. (2.13) and (2.14) in Eq. (2.12), we get the expression in the latter as

$$-\frac{K}{N}n_1\frac{\partial}{\partial\theta_1}\left\{\left[\int d\theta_2 f(\theta_1 - \theta_2)\rho_{1,0,...,0}(\theta_2, t)\right]\rho_{1,0,...,0}(\theta_1, t)\right\}$$

$$-\frac{K}{N}\sum_{r=1}^{p-1}n_{r+1}\frac{\partial}{\partial\theta_1}\left\{\left[\int d\theta_{N_r+1} f(\theta_1 - \theta_{N_r+1})\rho_{0,...,1,...,0}(\theta_{N_r+1}, t)\right]\rho_{1,0,...,0}(\theta_1, t)\right\}. \tag{2.15}$$

We now adopt a more convenient notation for one-body distribution functions that are the only distribution functions appearing in the last expression. More precisely, instead of using the subscripts that from now on we do not write any more, we specify the natural frequency to which the subscripts refer. The last expression then becomes

$$-\frac{K}{N}n_1\frac{\partial}{\partial\theta_1}\left\{\left[\int d\theta_2 f(\theta_1 - \theta_2)\rho(\theta_2, \omega_1, t)\right]\rho(\theta_1, \omega_1, t)\right\}$$

$$-\frac{K}{N}\sum_{r=1}^{p-1}n_{r+1}\frac{\partial}{\partial\theta_1}\left\{\left[\int d\theta_2 f(\theta_1 - \theta_{N_r+1})\rho(\theta_2, \omega_{n_r+1}, t)\right]\rho(\theta_1, \omega_1, t)\right\}. \tag{2.16}$$

We have almost arrived at the final expression. Note that the fraction n_r/N of oscillators with natural frequency equal to ω_r, in the limit $N \to \infty$ will be equal to the probability of occurrence of that frequency. Then, we may extend the last expression to the case of a continuum of frequencies distributed according to $g(\omega)$, by writing it as

$$-K\frac{\partial}{\partial\theta_1}\left\{\left[\int d\omega\, g(\omega)\int d\theta_2 f(\theta_1 - \theta_2)\rho(\theta_2, \omega, t)\right]\rho(\theta_1, \omega_1, t)\right\}. \tag{2.17}$$

The final expression, giving the Fokker-Planck equation for the one-body distribution function, is then obtained by substituting the last expression in Eq. (2.9), thus leading to

$$\frac{\partial}{\partial t}\rho(\theta,\omega,t) = -\omega\frac{\partial}{\partial \theta}\rho(\theta,\omega,t) + D\frac{\partial^2}{\partial \theta^2}\rho(\theta,\omega,t)$$

$$-K\frac{\partial}{\partial \theta}\left\{\left[\int d\omega' g(\omega') \int d\theta' f(\theta - \theta')\rho(\theta',\omega',t)\right]\rho(\theta,\omega,t)\right\}, \quad (2.18)$$

where we have removed the subscript attached to the variables θ and ω. The above is actually a set of Fokker-Planck equations, one for each ω, which are coupled through the interaction term. For each ω, the distribution $\rho(\theta,\omega,t)$ is normalized at all times (note that normalization is conserved by the Fokker-Planck equation), as

$$\int_0^{2\pi} d\theta \ \rho(\theta,\omega,t) = 1. \quad (2.19)$$

2.2 The Kuramoto Model

We now begin our study of the synchronization of coupled oscillators subject to noise, by focusing on the model that is by far the most celebrated one: the Kuramoto model. It was introduced in 1975 in its original noiseless version, and we have already considered it in Sect. 1.7 that was dedicated to the exposition of some results of noiseless systems. In the Kuramoto model, the interaction term, e.g., in the equations of motion (2.1), is given by posing $f(\theta) = -\sin\theta$. It is the simplest odd periodic function, i.e., such that $f(-\theta) = -f(\theta)$. The most general function should be a Fourier series with all possible odd terms: $\sin\theta$, $\sin(2\theta)$, $\sin(3\theta)$, ... However, study of the synchronization transition in the general case is considerably more difficult than for the Kuramoto model. As a result, the vast majority of studies reported in the literature have focussed on the simple $\sin\theta$ interaction. There are however some important works for the general case, and we will give some information in this regard towards the end of this chapter.

It is convenient for later reference to write explicitly the equations of motion (2.1) for the Kuramoto model, as [3]

$$\frac{d\theta_i}{dt} = \omega_i - \frac{K}{N}\sum_{j=1}^{N}\sin(\theta_i - \theta_j) + \eta_i(t), \quad i = 1, 2, \ldots, N, \quad (2.20)$$

where the prime in the summation that was there in Eq. (2.1) is no more necessary, since the sine term vanishes anyways for $i = j$. Let us also write down the Fokker-Planck equation (2.18) with $f(\theta) = -\sin\theta$, which now reads

$$\frac{\partial}{\partial t}\rho(\theta,\omega,t) = -\omega\frac{\partial}{\partial \theta}\rho(\theta,\omega,t) + D\frac{\partial^2}{\partial \theta^2}\rho(\theta,\omega,t)$$

$$+K\frac{\partial}{\partial \theta}\left\{\left[\int d\omega' g(\omega') \int d\theta' \sin(\theta - \theta')\rho(\theta',\omega',t)\right]\rho(\theta,\omega,t)\right\}. \quad (2.21)$$

As in Sect. 1.7, we adopt here the use of the order parameter r, which is the basic variable for the description of the synchronization transition:

$$r(t)e^{i\psi(t)} = \int d\omega \int_0^{2\pi} d\theta \, g(\omega)e^{i\theta}\rho(\theta, \omega, t). \tag{2.22}$$

Here, we have explicitly indicated the time dependence of $r(t)$ and its phase $\psi(t)$. More precisely, these two quantities are functionals of the distribution $\rho(\theta, \omega, t)$. With this definition, we have

$$\int d\omega' g(\omega') \int d\theta' \sin(\theta - \theta')\rho(\theta', \omega', t) = -r(t)\sin(\psi(t) - \theta), \tag{2.23}$$

so that we may write down our Fokker-Planck equation as

$$\frac{\partial}{\partial t}\rho(\theta, \omega, t) = -\frac{\partial}{\partial \theta}\left[(\omega + Kr(t)\sin(\psi(t) - \theta))\,\rho(\theta, \omega, t)\right] + D\frac{\partial^2}{\partial \theta^2}\rho(\theta, \omega, t). \tag{2.24}$$

The nonlinear nature of this equation is now hidden in the dependence of $r(t)$ and $\psi(t)$ on the function ρ itself. The solution will depend on $r(t)$ and $\psi(t)$, and hence, has to satisfy Eq. (2.22), which therefore becomes a self-consistent equation.

We will not be concerned with the general time-dependent solutions of the Fokker-Planck equation, but only with its stationary solutions. To be more precise about this point, in particular, to give a reasonable physical meaning to the search for stationary solutions, let us stress the following things.

The first point concerns the existence itself of one or many stationary solutions of Eq. (2.24). We remind the reader that in Chap. 1, while deriving the representation of the dynamics of interacting oscillators in terms of equations of motion containing only the phases, as in Eq. (2.1), we have assumed that the range of frequencies constituting the support of the distribution $g(\omega)$ does not include the origin and that its size is sufficiently small with respect to the value of the frequencies in its support.[2] These assumptions do not allow in general for the occurrence of time-independent solutions. To see this, let us analyze the simple case in which $g(\omega) = \delta(\omega - \omega_0)$, so that we need to consider Eq. (2.24) for only $\omega = \omega_0$. It is more convenient to make a change of variable: $\theta \to \widetilde{\theta} = \theta - \omega_0 t$. In this new variable, the Fokker-Planck equation (2.24) becomes

$$\frac{\partial}{\partial t}\rho(\widetilde{\theta}, \omega_0, t) = -\frac{\partial}{\partial \widetilde{\theta}}\left[Kr(t)\sin(\psi(t) - \widetilde{\theta} - \omega_0 t)\rho(\widetilde{\theta}, \omega_0, t)\right] + D\frac{\partial^2}{\partial \widetilde{\theta}^2}\rho(\widetilde{\theta}, \omega_0, t). \tag{2.25}$$

---

[2]In the following, we will often consider distributions $g(\omega)$ with a non-compact support, which is incompatible with these assumptions. However, while the choice of a non-compact support may give some mathematical advantages, the important thing from the physical point of view is that the mentioned assumptions hold when we consider only the frequencies where $g(\omega)$ is not negligibly small.

Before proceeding, let us remark that the change of variable corresponds to the passage in the point of observation, from the "laboratory" frame to a frame rotating with frequency ω_0 with respect to the laboratory frame. We now search for a solution of the last equation for which $\psi(t) = \omega_0 t$. It is then not difficult to see that the normalized solution is

$$\rho(\widetilde{\theta}, \omega_0, t) = \frac{e^{\frac{Kr}{D}\cos\widetilde{\theta}}}{\int_0^{2\pi} d\theta' e^{\frac{Kr}{D}\cos\theta'}}, \tag{2.26}$$

which is actually time independent. In the original variable θ, the solution reads

$$\rho(\theta, \omega_0, t) = \frac{e^{\frac{Kr}{D}\cos(\theta-\omega_0 t)}}{\int_0^{2\pi} d\theta' e^{\frac{Kr}{D}\cos\theta'}}. \tag{2.27}$$

We now have to satisfy the self-consistent equation (2.22), which in particular has to confirm that $\psi(t) = \omega_0 t$. The latter is effectively verified, while $r(t)$ is time-independent and has to satisfy

$$r = \frac{\int_0^{2\pi} d\theta \, \cos\theta \, e^{\frac{Kr}{D}\cos\theta}}{\int_0^{2\pi} d\theta \, e^{\frac{Kr}{D}\cos\theta}}. \tag{2.28}$$

Of course, the solution (2.27) exists only if this last equation admits a solution for positive r. We conclude that a stationary nontrivial solution may exist in the rotating frame but surely does not exist in the laboratory frame. However, we do also have a stationary solution, represented by the homogeneous distribution $\rho(\theta, \omega_0) = 1/(2\pi)$, which corresponds to the case with $r = 0$. The existence of more than one stationary state (in the $\widetilde{\theta}$ variable), the nontrivial one and the homogeneous one, confirms what we have discussed in Chap. 1, namely, that for our nonlinear Fokker-Planck equation, the uniqueness of the stationary solution is not guaranteed. From the foregoing discussion, we may expect that also for general $g(\omega)$, nontrivial stationary solutions do not exist. In spite of this, we will see that with a change of variable similar to the one adopted for our simple example, we can obtain stationary solutions.

The second point concerns the approach at long times of the time-dependent solution of the Fokker-Planck equation to one of the stationary states. As underlined in Chap. 1, since the drift coefficient in Eq. (2.24) depends on the function ρ, the conditions of the theorem that guarantees this approach are not verified. Actually, for the above example with a single frequency ω_0, a sort of H-theorem, known from equilibrium statistical physics [2], exists. However, it cannot be extended to the case of a general $g(\omega)$; see Appendix 1. In spite of this fact, the system of oscillators converges in general to a stationary state (in terms of suitable variables), as may also be confirmed by numerical simulations.

In the following section, we study the Kuramoto model in its most-considered setting, i.e., in which the frequency distribution $g(\omega)$ is unimodal and symmetric with respect to the frequency value at which the distribution attains its maximum.

2.3 Unimodal Symmetric $g(\omega)$

We study here the Kuramoto model in the case in which the frequency distribution $g(\omega)$ has the following properties: it has a single maximum at $\omega = \Omega$ with respect to which it is symmetric. The latter fact may be expressed as $g(\Omega + \omega_1) \geq g(\Omega + \omega_2)$ if $|\omega_1| \leq |\omega_2|$, together with $g(\Omega + \omega) = g(\Omega - \omega)$. These properties imply that Ω is the average frequency of the distribution. We have already considered this case in Chap. 1 while analyzing the noiseless Kuramoto model. Let us go to a moving frame that rotates with frequency Ω with respect to an inertial frame. Defining $\tilde{\theta} = \theta - \Omega t, \tilde{\psi}(t) = \psi(t) - \Omega t$ and $\tilde{\omega} = \omega - \Omega$, the Fokker-Planck equation (2.24) when expressed in these new variables reads

$$\frac{\partial}{\partial t} \rho(\tilde{\theta}, \Omega + \tilde{\omega}, t) = -\frac{\partial}{\partial \tilde{\theta}} \left[\left(\tilde{\omega} + Kr(t) \sin(\tilde{\psi}(t) - \tilde{\theta}) \right) \rho(\tilde{\theta}, \Omega + \tilde{\omega}, t) \right]$$

$$+ D \frac{\partial^2}{\partial \tilde{\theta}^2} \rho(\tilde{\theta}, \Omega + \tilde{\omega}, t). \tag{2.29}$$

From now on (except in Appendix 1), we drop for convenience the tilde for the quantities defined in the rotating frame, but the reader must remember that we are working in such a frame. We also drop the indication of the average frequency Ω in the distribution ρ. Hence, the Fokker-Planck equation may be written as

$$\frac{\partial}{\partial t} \rho(\theta, \omega, t) = -\frac{\partial}{\partial \theta} \left[(\omega + Kr(t) \sin(\psi(t) - \theta)) \rho(\theta, \omega, t) \right] + D \frac{\partial^2}{\partial \theta^2} \rho(\theta, \omega, t). \tag{2.30}$$

The frequency distribution $g(\omega)$ is now unimodal and symmetric: $g(\omega_1) \geq g(\omega_2)$ if $|\omega_1| \leq |\omega_2|$, and $g(\omega) = g(-\omega)$.

We now assume that there is a stationary solution of the Fokker-Planck equation, which therefore implies that both $r(t)$ and $\psi(t)$ are time independent. Redefining if necessary the origin of the angle θ, we may set the time-independent value of ψ to zero without any loss of generality, and are therefore led to consider the equation

$$-\frac{\partial}{\partial \theta} \left[(\omega - Kr \sin \theta) \rho(\theta, \omega) \right] + D \frac{\partial^2}{\partial \theta^2} \rho(\theta, \omega) = 0. \tag{2.31}$$

The above is an ordinary differential equation that may be integrated to obtain [3]

$$\rho(\theta, \omega; r) = C e^{\frac{Kr \cos \theta + \omega \theta}{D}} \left[1 + \left(e^{-\frac{2\pi \omega}{D}} - 1 \right) \frac{\int_0^\theta d\theta' e^{-\frac{Kr \cos \theta' + \omega \theta'}{D}}}{\int_0^{2\pi} d\theta' e^{-\frac{Kr \cos \theta' + \omega \theta'}{D}}} \right], \tag{2.32}$$

where $C = \rho(0, \omega; r) \exp(-Kr/D)$ is the normalization constant. In the above equation, we have indicated the parametric dependence of ρ on the order parameter r that has to be determined self-consistently. It may be verified that the function $\rho(\theta, \omega; r)$ is periodic in θ, as it should be, i.e., $\rho(\theta + 2\pi, \omega; r) = \rho(\theta, \omega; r)$ for each θ.

For the solution (2.31) to be acceptable, the self-consistent equation

$$r = \int d\omega \int_0^{2\pi} d\theta \; g(\omega) e^{i\theta} \rho(\theta, \omega; r) \tag{2.33}$$

must be satisfied. The imaginary part of the right hand side vanishes because of the property that we have $\rho(-\theta, -\omega; r) = \rho(\theta, \omega; r)$, see Eq. (2.31). Then, we are left with the self-consistent equation

$$r = \int d\omega \int_0^{2\pi} d\theta \; g(\omega) \cos\theta \rho(\theta, \omega; r). \tag{2.34}$$

We may easily see that the homogeneous distribution $\rho(\theta, \omega) = 1/(2\pi)$, for which $r = 0$, is always an acceptable solution of the above equation; in the following subsection, we will study its dynamical stability. On the other hand, the self-consistent equation (2.34) does not always admit a solution with $r > 0$.

We see that for $K = 0$, the only acceptable solution to the self-consistent equation is the homogeneous one. This is to be expected, since without any interaction, the oscillators will not have a tendency to synchronize with one another and form a state with a positive order parameter. In the study of the noiseless case in Chap. 1, we have seen that a solution with $r > 0$ exists only for K larger than a threshold value K_c, which bifurcates at K_c from the homogeneous state. On physical grounds, we may guess that for a given coupling constant K and a given frequency distribution $g(\omega)$, the value of the order parameter r of the stable stationary state decreases on increasing the constant D. Thus, we may expect that also for the noisy case, the solution with $r > 0$ exists only for K above a given threshold, and that this threshold increases with increase of D.

To find the bifurcation value of K, we study the solution (2.32) in the limit $r \to 0$, by making a power series expansion in r of the solution, and consequently, of the self-consistent equation (2.34). One may argue that the right hand side of Eq. (2.34) contains only odd powers of r. Performing the power series expansion and keeping only the first two terms, one arrives at the following expression:

$$
\begin{aligned}
r &= \frac{Kr}{2} \int d\omega \; g(\omega) \frac{D}{\omega^2 + D^2} - \frac{K^3 r^3}{4D} \int d\omega \; g(\omega) \left[\frac{1}{\omega^2 + 4D^2} - \frac{\omega^2}{(\omega^2 + D^2)^2} \right] \\
&= \frac{Kr}{2} \int d\omega \; g(D\omega) \frac{1}{\omega^2 + 1} - \frac{K^3 r^3}{4D^2} \int d\omega \; g(D\omega) \left[\frac{1}{\omega^2 + 4} - \frac{\omega^2}{(\omega^2 + 1)^2} \right], \tag{2.35}
\end{aligned}
$$

where in the second line, obtained from the first line with the trivial change of integration variable $\omega \to D\omega$, we have an expression that will prove somewhat more convenient for a study of the limit of small noise.

Apart from the trivial solution $r = 0$, Eq. (2.35) can admit another real solution with $r > 0$. We are going to prove that for a given D, this solution exists for K above a certain threshold. First, we determine this threshold that defines the value of K at which the solution $r > 0$ bifurcates from the trivial solution $r = 0$. The threshold K_c is the value of K for which the coefficient of r on the right hand side of Eq. (2.35) is equal to 1. We get

$$K_c = 2 \left[\int d\omega \, g(\omega) \frac{D}{\omega^2 + D^2} \right]^{-1} = 2 \left[\int d\omega \, g(D\omega) \frac{1}{\omega^2 + 1} \right]^{-1}. \tag{2.36}$$

Deriving the right hand side with respect to D shows that K_c increases with D, as one would have expected. However, this may not be true for non-unimodal frequency distributions, as we will see later.

Equation (2.35) has a positive solution for r for $K > K_c$. This may be seen as follows: The expression in square brackets in the second term on the right hand side vanishes when integrated with respect to ω. On the other hand, the expression itself is positive for $|\omega| < D/\sqrt{2}$ and negative otherwise. This means that the integral extended to $|\omega| > D/\sqrt{2}$ is negative, but equal, in absolute value, to the integral extended to $|\omega| < D/\sqrt{2}$. Then, multiplying the expression in square brackets for a unimodal and symmetric $g(\omega)$, the integral extended to $|\omega| > D/\sqrt{2}$ attains a negative value that in modulus is smaller than the integral extended to $|\omega| < D/\sqrt{2}$. We conclude that the integral in the second term on the right hand side of Eq. (2.35) is positive. As a consequence, a real solution in r exists only for $K > K_c$. Using the integral expressions in the second line of Eq. (2.35), the solution is given by

$$r = \frac{2D}{\sqrt{K^3}} \sqrt{\frac{\frac{K}{2} \int d\omega g(D\omega) \frac{1}{\omega^2 + 1} - 1}{\int d\omega g(D\omega) \left[\frac{1}{\omega^2 + 4} - \frac{\omega^2}{(\omega^2 + 1)^2} \right]}}. \tag{2.37}$$

We remind that this expression may be used for small values of r, because it has been derived from Eq. (2.35), which in turn is valid for small r as it contains only the first two terms of a power series expansion in r. For $K = K_c + \Delta K$ with $0 < \Delta K \ll K_c$, which is the range of its validity, the above expression may be written at first order in ΔK as

$$r = \frac{2D}{K_c^2} \sqrt{\frac{\Delta K}{\int d\omega g(D\omega) \left[\frac{1}{\omega^2 + 4} - \frac{\omega^2}{(\omega^2 + 1)^2} \right]}}. \tag{2.38}$$

It is interesting to see what happens in the limit of vanishing noise, i.e., as $D \to 0$. First, by posing $D = 0$ in Eq. (2.36), we obtain $K_c = 2/(\pi g(0))$, which coincides with Eq. (1.51). Second, a power expansion in D of the denominator of the square root in Eq. (2.38) gives $-D^2 \pi g''(0)/4$. Thus, in the limit $D \to 0$, we obtain

$$r = \frac{4}{K_c^2} \sqrt{-\frac{\Delta K}{\pi g''(0)}}, \tag{2.39}$$

which coincides with Eq. (1.56).

On the basis of the above discussions, we thus conclude that a continuous transition is obtained on increasing the coupling constant K for a given value of D, from the incoherent solution with $r = 0$ at low-K values to a state with $r > 0$ at high-K values. The reader could object that the incoherent solution with $r = 0$ exists for any value of K, as we have remarked after Eq. (2.34). However, we will shortly see that for $K > K_c$, such a solution is dynamically unstable, leaving the state with $r > 0$ as the only stable one.

Before proceeding to a study of the dynamical stability of the $r = 0$ solution, we analyze the synchronization properties of the system when one has $r > 0$. In Chap. 1, we have seen that the phase locking that can occur in the noiseless case is expected to be only approximate when there is noise: the phases of the different oscillators are never strictly locked to have a constant difference due to the presence of noise. Nevertheless, synchronization occurs in an average sense, thus still giving rise to a positive r. A perfect synchronization of an oscillator would require it to have zero frequency and to have come to a standstill in the rotating frame in which we are working. Its actual (average) frequency in the state with $r > 0$ may be obtained from the time-independent Fokker-Planck equation (2.31) by plugging its normalized solution (2.32). In fact, from the theory of the Fokker-Planck equation, we know that it may be written in the form [1]

$$\frac{\partial}{\partial t} \rho(\theta, \omega, t) = -\frac{\partial}{\partial \theta} S(\theta, \omega, t), \tag{2.40}$$

where $S(\theta, \omega, t)$ is the probability current. In the stationary state, the current is then independent of θ and t. From Eq. (2.31), we see that the current is given by

$$S(\omega) = (\omega - Kr \sin \theta)\, \rho(\theta, \omega) - D\frac{\partial}{\partial \theta}\rho(\theta, \omega), \tag{2.41}$$

with $\rho(\theta, \omega)$ given by Eq. (2.32). This expression for the current is of course valid also in the $r = 0$ homogeneous state with $\rho(\theta, \omega) = 1/(2\pi)$. In this case, we have $S = \omega/(2\pi)$. This is what we would have expected, since for $r = 0$, each oscillator makes independent uniform rotation at its own natural frequency. For the general case with $r > 0$, we cannot write an expression simpler than Eq. (2.41). However, we may have an idea of the average frequency of each oscillator for small r, by resorting to a power series expansion in r of Eq. (2.41). We obtain to lowest order in r the result

$$S(\omega) = \frac{1}{2\pi} \left[\omega - \frac{K^2 r^2}{2} \frac{\omega}{\omega^2 + D^2} \right]. \tag{2.42}$$

Then, in the partially coherent state with $r > 0$ (but small), the oscillators with natural frequency ω acquire an average frequency

$$\omega' = \omega - \frac{K^2 r^2}{2} \frac{\omega}{\omega^2 + D^2}. \tag{2.43}$$

This expression does not vanish for[3] $\omega \neq 0$. Thus, there is no oscillator that acquires a vanishing frequency, i.e., no oscillator is perfectly phase locked, although the shift is towards having synchronization, since $|\omega|$ decreases, so that ω' is closer to zero than ω. This situation has to be compared with that occurring in the noiseless case, where we have seen that the oscillators with natural frequency $\omega < Kr$ are perfectly locked, while the remaining ones continue to drift. However, also the drifting oscillators feel some effects. In fact, from the integration of Eq. (1.43) for $\omega > Kr$, we find that the motion of such oscillators is periodic with a period equal to $2\pi/\sqrt{\omega^2 - (Kr)^2}$. Thus, their frequency is shifted to $\omega' = \sqrt{\omega^2 - (Kr)^2}$. We note that the power series expansion of the last expression gives at lowest order in r the expression in Eq. (2.43) for $D = 0$.

2.3.1 The Stability of the Incoherent State

The homogeneous stationary state $\rho(\theta, \omega) = 1/(2\pi)$ exists for any value of the coupling constant K. However, it is dynamically unstable for $K > K_c$. The stability analysis is performed by studying the linearized Fokker-Planck equation that is obtained by expanding $\rho(\theta, \omega, t)$ as

$$\rho(\theta, \omega, t) = \frac{1}{2\pi} + \delta\rho(\theta, \omega, t), \tag{2.44}$$

with $\delta\rho \ll 1$, substituting in Eq. (2.21), and keeping only the terms that are at most linear in $\delta\rho$. We then obtain the linear equation

$$\frac{\partial}{\partial t}\delta\rho(\theta, \omega, t) = -\omega\frac{\partial}{\partial\theta}\delta\rho(\theta, \omega, t) + D\frac{\partial^2}{\partial\theta^2}\delta\rho(\theta, \omega, t)$$
$$+ \frac{K}{2\pi}\int d\omega' g(\omega')\int d\theta' \cos(\theta - \theta')\delta\rho(\theta', \omega', t). \tag{2.45}$$

To study the last equation, we make a Fourier expansion of $\delta\rho$:

$$\delta\rho(\theta, \omega, t) = \sum_{k=-\infty}^{+\infty} \widehat{\delta\rho}_k(\omega, t)e^{ik\theta}. \tag{2.46}$$

Substituting this expansion in Eq. (2.45), we get the following equation for the k-th Fourier component:

---

[3] Actually, it vanishes for $\omega^2 = (Kr)^2/2 - D^2$, if this expression is positive. However, this requires $(Kr)/D$ to be of order 1, making the power series expansion in r, which is in $(Kr)/D$, no longer suitable: The power series expansion at the lowest order is good for $(Kr)/D$ small with respect to 1.

$$\frac{\partial}{\partial t}\widehat{\delta\rho}_k(\omega, t) = -ik\omega\widehat{\delta\rho}_k(\omega, t) - Dk^2\widehat{\delta\rho}_k(\omega, t)$$

$$+ \frac{K}{2}\left(\delta_{k,1} + \delta_{k,-1}\right)\int d\omega' g(\omega')\widehat{\delta\rho}_k(\omega', t), \qquad (2.47)$$

where $\delta_{i,j}$ is the Kronecker delta symbol. For $k \neq \pm 1$, there is no contribution from the integral term, and we have

$$\frac{\partial}{\partial t}\widehat{\delta\rho}_k(\omega, t) = -ik\omega\widehat{\delta\rho}_k(\omega, t) - Dk^2\widehat{\delta\rho}_k(\omega, t), \qquad (2.48)$$

which shows that for $k \neq \pm 1$, there is an exponential decay of $\widehat{\delta\rho}_k(\omega, t)$ with rate Dk^2. Actually, one has a continuous spectrum of stable modes,[4] identified by the values of ω in the support of $g(\omega)$.

For the study of the Fourier components with $k = \pm 1$, we pose

$$\widehat{\delta\rho}_{\pm 1}(\omega, t) = \widetilde{\delta\rho}_{\pm 1}(\omega, \lambda)e^{\lambda t}. \qquad (2.49)$$

Substituting in Eq. (2.47), we obtain

$$(\lambda \pm i\omega + D)\,\widetilde{\delta\rho}_{\pm 1}(\omega, \lambda) = \frac{K}{2}\int d\omega' g(\omega')\widetilde{\delta\rho}_{\pm 1}(\omega', \lambda). \qquad (2.50)$$

Also in this case, we have a continuous spectrum of stable modes, one for each ω value in the support of $g(\omega)$. If ω_0 is one such value, the corresponding stable mode for $\widetilde{\delta\rho}_{\pm 1}$ has $\lambda = -D \mp i\omega_0$, while $\widetilde{\delta\rho}_{\pm 1}$, normalized so that the right hand side of Eq. (2.50) is equal to 1 (since Eq. (2.50) is linear, we may choose the normalization that is more convenient), is given by

$$\widetilde{\delta\rho}_{\pm 1}(\omega, -D \mp i\omega_0) = \mp i\mathscr{P}\frac{1}{\omega - \omega_0} + c_{\pm 1}(\omega_0)\delta(\omega - \omega_0), \qquad (2.51)$$

where

$$c_{\pm 1}(\omega_0)g(\omega_0) = \frac{2}{K} \pm i\mathscr{P}\int d\omega'\frac{g(\omega')}{\omega' - \omega_0}, \qquad (2.52)$$

and where $\mathscr{P}$ indicates the principal value.[5]

---

[4]The Fourier component with $k = 0$ would be linearly neutrally stable in this framework. However, such a component vanishes identically due to the normalization of ρ: from Eq. (2.44), one finds that the integral in θ of $\delta\rho$ must vanish.

[5]From the mathematical point of view, Eqs. (2.48) and (2.50) may be considered also for ω outside the support of $g(\omega)$, i.e., for those values of ω, if any, for which $g(\omega) = 0$. In this case, nothing changes as regards the solution of Eq. (2.48), while the solution of Eq. (2.50) for $g(\omega_0) = 0$ is $\widetilde{\delta\rho}_{\pm 1}(\omega, \omega_0) = \delta(\omega - \omega_0)$ (in this case the right hand side of this equation cannot be made equal to 1).

On the other hand, there is also a discrete spectrum. The corresponding values of λ may be found by rewriting Eq. (2.50) as

$$\widetilde{\delta\rho}_{\pm 1}(\omega, \lambda) = \frac{K}{2\left(\lambda \pm i\omega + D\right)} \int d\omega' g(\omega') \widetilde{\delta\rho}_{\pm 1}(\omega', \lambda). \qquad (2.53)$$

By imposing again the normalization for which the right hand side of Eq. (2.50) is equal to 1, we obtain from the last equation the dispersion relation

$$\frac{K}{2} \int d\omega \frac{g(\omega)}{\lambda \pm i\omega + D} = 1. \qquad (2.54)$$

It is not difficult to see that the last equation can have only real solutions for λ. If λ_r and λ_i are the real and the imaginary part of λ, then the real and the imaginary parts of Eq. (2.54) give

$$\frac{K}{2} \int d\omega\, g(\omega) \frac{\lambda_r + D}{(\lambda_r + D)^2 + (\lambda_i \pm \omega)^2} = 1, \qquad (2.55)$$

$$\int d\omega\, g(\omega) \frac{\lambda_i \pm \omega}{(\lambda_r + D)^2 + (\lambda_i \pm \omega)^2} = 0. \qquad (2.56)$$

The second equation can be satisfied only for $\lambda_i = 0$; in fact, with the change of variable $\omega \pm \lambda_i = x$, it may be written as

$$\int dx\, g(x \mp \lambda_i) \frac{x}{(\lambda_r + D)^2 + x^2} = 0. \qquad (2.57)$$

It is a simple exercise to see that for $\lambda_i \neq 0$, the left hand side cannot be equal to zero, since $g(\omega)$ is symmetric and unimodular. On the other hand, for $\lambda_i = 0$, the equation is trivially satisfied. We are thus left with the equation (with $\lambda \equiv \lambda_r$)

$$\frac{K}{2} \int d\omega\, g(\omega) \frac{\lambda + D}{(\lambda + D)^2 + \omega^2} = 1, \qquad (2.58)$$

valid for both $k = 1$ and $k = -1$. The left hand side is positive for $\lambda \geq -D$, so that a solution in λ of this equation, if it exists, must satisfy this bound. In case such a solution does not exist, the uniform state is stable, since all the eigenvalues of the continuous spectrum have a negative real part equal to $-Dk^2$ for $k = 1, 2, \ldots$. Stability occurs also if Eq. (2.58) does have a solution, but this solution is negative; on the other hand, if the solution is positive, the uniform state is unstable.

It is not difficult to find the range of K for which the uniform state is unstable. With the change of variable $\omega = (\lambda + D)x$, Eq. (2.58) may be written for $\lambda \geq -D$ as

$$\frac{K}{2} \int dx\, g[(\lambda + D)x] \frac{1}{1 + x^2} = 1. \qquad (2.59)$$

The left hand side decreases with increasing λ, as may be seen by its derivative with respect to λ, given by

$$\frac{K}{2} \int dx \, g'[(\lambda + D)x] \frac{x}{1 + x^2}, \tag{2.60}$$

which is clearly negative; furthermore, it tends to 0 for $\lambda \to \infty$. Therefore, there can be at most one solution for Eq. (2.59); this solution is positive if the value of the left hand side is larger than 1 for $\lambda = 0$. Consequently, the uniform state is unstable when K is larger than

$$K_c = 2 \left[\int dx \, g(Dx) \frac{1}{1 + x^2} \right]^{-1}. \tag{2.61}$$

We note that this is exactly the same threshold found above for the existence of the state with $r > 0$. Thus, as soon as the $r > 0$ state exists, the incoherent uniform state becomes unstable.

2.3.1.1   The Noiseless Limit

The result that the threshold for the existence of the state with $r > 0$ coincides with the instability threshold of the uniform state is valid also in the noiseless limit $D \to 0$. In that case, we have $K_c = 2/(\pi g(0))$. However, we note a peculiarity: For $K < K_c$, there is no eigenvalue with a positive real part, and in fact, all the eigenvalues have a vanishing real part: the real part, equal to $-Dk^2$, goes to 0 as $D \to 0$. Then, we conclude that the uniform state is never linearly stable, but at most neutrally stable. However, it may be proved in spite of this that for $K < K_c$, a state that initially has $r > 0$ decays to the uniform state. The interested reader may find a proof of this fact in the brief review by Strogatz [4], where it is put in evidence that this behavior is mathematically similar to the Landau damping in plasmas.[6]

2.3.2   Examples of $r(K)$: Computations and Simulations

We now consider in more detail two examples of the distribution $g(\omega)$, namely, a Gaussian and a Lorentzian. Let us make some general considerations about the form of the function $r(K)$. Firstly, we note the following point: Studying the solution of the self-consistent equation (2.34) for K close to the threshold K_c, we have seen by using the power expansion in Eq. (2.35) and obtaining Eq. (2.38) that a solution with (small) positive r exists for K larger than K_c, but not for K smaller than K_c (in the theory of bifurcations, this phenomenon is called a supercritical transition [5]).

---

[6]The mathematically oriented reader will also appreciate that the tendency to 0 of r has to be intended in the weak topology.

Considering the function $r(K)$ for general K, and not just for K close to K_c, we expect a behavior of the type sketched with the full line in Fig. 2.1. At $K = K_c$, the curve starts with an infinite derivative, in agreement with Eq. (2.38) that shows that for small $K - K_c$, the quantity r is proportional to $(K - K_c)^{1/2}$. Then, by increasing K, we have a monotonic increase of r that tends to 1 as $K \to \infty$ (the negative second derivative in the full range of this illustrative curve is used only for simplicity). The approach of r to unity at large K may be argued on physical grounds to occur for any distribution $g(\omega)$, and it may be proved by studying the large K limit of Eq. (2.32). On the other hand, for symmetric bimodal frequency distributions, we may expect other types of $r(K)$ curves, as, e.g., the dashed line in Fig. 2.1. In fact, now that the denominator of the term under square root in Eq. (2.38) can be negative, a solution can exist for K smaller than K_c. In the noiseless limit, this is what surely occurs, since in Eq. (2.39), the term $g''(0)$ is positive. Although for a symmetric unimodal distribution, the dashed line cannot occur as the power series expansion (2.38) has a solution only for $K > K_c$, one could however envisage functions of the forms depicted with the dot-dashed and the dotted line in Fig. 2.1. These curves satisfy the requirement that for K close to K_c a small positive r occurs only for K larger than K_c, but have properties that are different from those of the full line. In particular, the dot-dashed curve has a range of K where, as for the dashed curve, there is more than one possible r for a given K; in principle, the re-entrance of the dot-dashed line could be even more pronounced than shown in the figure, for example, it may extend to the region $K < K_c$.

On the other hand, the dotted line has one value of r for each $K > K_c$, but $r(K)$ does not increase monotonically with K. In the noiseless case, we have proved in Chap. 1 that the function $r(K)$ is of the form represented by the full line, since there

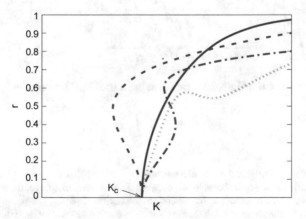

Fig. 2.1 Full line: sketch of the guessed form of the function $r(K)$ for a symmetric unimodal frequency distribution. Dashed line: possible form of $r(K)$ for a symmetric bimodal frequency distribution; for K close to K_c the function starts towards the left (subcritical bifurcation). Dot-dashed line and dotted line: other possible forms for a symmetric unimodal frequency distribution; in the text it is guessed that these forms never occur (see also Appendix 2)

is only one possible value of positive r for each $K > K_c$. We do not have a similar proof here for the noisy case.[7]

Nevertheless, we expect that, also in the noisy case, for symmetric unimodal distributions, there is only one positive solution for r for $K > K_c$, with r increasing monotonically with increasing K. Then the form of the function $r(K)$ is expected to be always the one represented by the full line in Fig. 2.1. In Appendix 2, a justification of this fact, although not a real proof, is given. The argument presented makes use of some of the results described in the next subsection dedicated to the study of a uniform frequency distribution $g(\omega)$; therefore, before reading Appendix 2, it is useful to go through the next subsection.

Now we turn to discussing results obtained for two particular distributions, a Gaussian and a Lorentzian. To this end, we consider the following frequency distribution functions:

$$g(\omega) = \frac{1}{\sqrt{2\pi\sigma^2}} e^{-\frac{\omega^2}{2\sigma^2}}, \tag{2.62}$$

$$g(\omega) = \frac{\sigma}{\pi} \frac{1}{\omega^2 + \sigma^2}. \tag{2.63}$$

The parameter σ characterizes the width of the distributions. While for the Gaussian σ is equal to the standard deviation, i.e., to the square root of the expectation value of ω^2, this is not the case for the Lorentzian that has an infinite standard deviation.

As a first step of our analysis, we compute the threshold value K_c of the coupling constant by using Eq. (2.36). The corresponding integrals may be computed in closed form, since we have

$$\frac{1}{\sqrt{2\pi\sigma^2}} \int_{-\infty}^{+\infty} d\omega\, e^{-\frac{\omega^2}{2\sigma^2}} \frac{D}{\omega^2 + D^2} = \sqrt{\frac{\pi}{2\sigma^2}}\, e^{\frac{D^2}{2\sigma^2}} \left[1 - \mathrm{erf}\left(\frac{D}{\sqrt{2\sigma^2}}\right)\right], \tag{2.64}$$

$$\frac{\sigma}{\pi} \int_{-\infty}^{+\infty} d\omega \frac{1}{\omega^2 + \sigma^2} \frac{D}{\omega^2 + D^2} = \frac{1}{\sigma + D}, \tag{2.65}$$

where in the case of the Gaussian, we have used the error function $\mathrm{erf}(x)$ defined by

$$\mathrm{erf}(x) \equiv \frac{2}{\sqrt{\pi}} \int_0^x dy\, e^{-y^2}. \tag{2.66}$$

---

[7]Earlier we have proved that in the case of a unimodal symmetric distribution, the incoherent state is dynamically stable for $K < K_c$, but also this does not disallow in principle for the occurrence of the dot-dashed curve with a re-entrance that extends to the region $K < K_c$. The situation is analogous to the existence of metastable states in thermodynamics. Such states realize local minima of the free energy, and a sufficiently small perturbation does not destroy them; however, the global minimum of the free energy occurs for the equilibrium state, and the system when given sufficient time for evolution leaves the metastable state to eventually reach the equilibrium state. As we have already remarked, our systems do not possess a free energy, and such evaluations cannot be done; the linear stability analysis can only tell us if a state is dynamically stable or not. Even if stable, there can be other stable states that the system prefers, analogously to the preference of the equilibrium state with respect to the metastable state. We will come back to this point in Chap. 3.

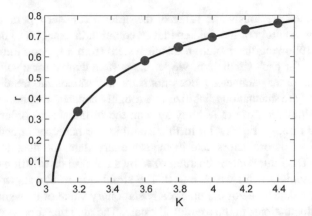

Fig. 2.2 Full line: the function $r(K)$ for a zero mean and unit width ($\sigma = 1$) Gaussian frequency distribution and for $D = 1$, obtained by numerically solving the self-consistent equation (2.34). Circles: values of $r(K)$ obtained from the numerical simulation of the equations of motion (2.20). The computation in the latter case is explained in the text

The function $r(K)$ is obtained by numerically solving the self-consistent equation (2.34); this function will further depend on D and on σ. From the structure of the function $\rho(\theta, \omega; r)$ given in Eq. (2.32), it is clear that the dependence will be through K/D and σ/D; therefore, we do not loose generality by studying the case with $D = 1$. In our example, we also make the choice $\sigma = 1$; the structure of the function $r(K)$ does not change on choosing other values of σ. Using Eqs. (2.64) and (2.65), after posing $D = 1$ and $\sigma = 1$, in Eq. (2.36), we then obtain the following values for the threshold coupling constant K_c:

$$K_c = \sqrt{\frac{8}{\pi e}} \frac{1}{1 - \mathrm{erf}\left(\frac{1}{\sqrt{2}}\right)} \approx 3.0503; \quad \text{(Gaussian)}, \tag{2.67}$$

$$K_c = \frac{2}{\frac{1}{2}} = 4; \quad \text{(Lorentzian)}. \tag{2.68}$$

In Fig. 2.2, we show the function $r(K)$ for the Gaussian frequency distribution function, obtained by solving numerically Eq. (2.34). Appendix 3 gives the details of how this numerical solution is obtained. In the same plot, we show the value of r obtained in numerical simulations of the equations of motion (2.20) for some values of K. The equations of motion are of the Langevin type, and contain a stochastic term, related to the Gaussian white noise. There is a standard procedure for the numerical integration of these stochastic equations; we prefer to defer a brief explanation of this procedure to Chap. 3, where it is applied to the case of oscillators with inertia with a dynamics governed by equations of motion that are second order in time. The procedure for first-order dynamics may be simply deduced from the more-general second-order case in Chap. 3.

As explained previously, the value of the order parameter r pertaining to a stable stationary solution of the Fokker-Planck equation is expected to be achieved at large times following the evolution of the system from a generic initial condition. Accordingly, for each simulation, we let the system evolve for a sufficiently long time until the order parameter r does not show any monotonic trend of change in its values and the fluctuations stabilize to a roughly constant amplitude. We remind that in the simulation of the N-body system, the instantaneous order parameter r is computed through Eq. (1.41). In this example, we have performed simulations with $N = 5 \times 10^5$ oscillators, and in the stationary state, the fluctuations of r are quite small. The initial state is characterized by a uniform distribution of the phases θ_i's of the oscillators, for which $r \approx 0$. The simulation results shown with circles in Fig. 2.2 give the value of the practically stationary value of r, averaged over the very small fluctuations, in the asymptotic state. The agreement between the results obtained from simulations and from the stationary states of the Fokker-Planck equation is quite good. Not reported here are simulation results for the case $K < K_c$ that confirm a vanishing r.

Going now to the Lorentzian frequency distribution, Fig. 2.3 shows the function $r(K)$ obtained by solving the self-consistent equation (2.34), together with the results of the numerical simulations of the equations of motion (2.20). The number of oscillators used in the simulations is again $N = 5 \times 10^5$, and the criterion to evaluate the stationary order parameter r is the same as for the Gaussian distribution reported above. From the figure, it may be seen that the agreement for the three smaller simulated values of K is not extremely good, although quite satisfying (the relative difference is about 3% for $K = 4.1$, about 2% for $K = 4.3$, and about 1% for $K = 4.5$). This small disagreement could be due to the relative weight of the very large frequencies in dictating the dynamics, which is more important

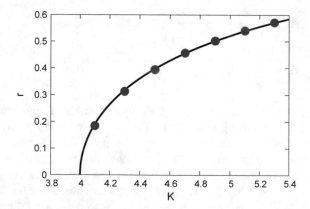

Fig. 2.3 Full line: the function $r(K)$ for a Lorentzian frequency distribution with $\sigma = 1$ and for $D = 1$, obtained by numerically solving the self-consistent equation (2.34). Circles: values of $r(K)$ obtained from the numerical simulation of the equations of motion (2.20). The computation in the latter case is explained in the text

for the Lorentzian than for the Gaussian: their contribution could not be evaluated perfectly in the numerical solution of the self-consistent equation (see also Appendix 3), while in the simulations, their approach to the stationary state could be very slow and not yet perfectly achieved during the time of the simulation runs. In spite of these technical subtleties, we are fully justified to conclude that also for the Lorentzian, the simulation results agree very well with the evaluations of r obtained from the Fokker-Planck equation.

We conclude this section with the following remark. We have proved that for $K < K_c$, the uniform state is stable, while for $K > K_c$, it is unstable; we have also shown that for $K > K_c$, the system approaches the partially synchronized state with $r > 0$; this is a very strong evidence that the latter state is stable. However, we must be aware that mathematically, the stability of this state has not been proved, although we are ready to accept it on physical grounds.

2.3.3 Uniform $g(\omega)$

The case of a uniform distribution of frequencies is very interesting. The function $g(\omega)$ is equal to a constant between $\omega = -\omega_1$ and $\omega = +\omega_1$, where ω_1 is a given positive number, and vanishes elsewhere; for reasons of normalization, the uniform value of $g(\omega)$ where it does not vanish is equal to $1/(2\omega_1)$. Equation (2.39) shows that the power expansion that should give r for K close to K_c does not work in the noiseless case for an uniform distribution, since $g''(0)$ for the latter vanishes. But before dealing with this fact, we consider the general noisy case. We may take $D = 1$ as before without loss of generality.

We start our analysis by computing the threshold value of the coupling constant, K_c. From Eq. (2.36), we obtain

$$K_c = 2 \left[\frac{1}{2\omega_1} \int_{-\omega_1}^{+\omega_1} d\omega \frac{1}{\omega^2 + 1} \right]^{-1} = \frac{2\omega_1}{\arctan \omega_1}. \tag{2.69}$$

For our numerical study, we choose the value $\omega_1 = 0.5$. The above equation then gives $K_c = 1/(\arctan 0.5) \approx 2.1568$. The numerical solution of the self-consistent equation (2.34) gives the function $r(K)$ shown in Fig. 2.4, where we also show the results obtained from numerical simulations of the equations of motion of a system with $N = 5 \times 10^5$ oscillators and for chosen values of K. As in the Gaussian and the Lorentzian case discussed above, here too we observe a very good agreement between the results of the N-body simulations and of the Fokker-Planck equation. Furthermore, qualitatively, the behavior of the system is not different from the cases with the Gaussian and the Lorentzian frequency distribution: at a certain threshold value K_c, there is a continuous (or, second-order) synchronization transition from a state with vanishing order parameter to a state with $r > 0$. However, we now show that this situation changes in the limit of vanishing noise.

Fig. 2.4 Full line: the
function $r(K)$ for a uniform
frequency distribution with
$\omega_1 = 0.5$ and for $D = 1$,
obtained by numerically
solving the self-consistent
equation (2.34). Circles:
values of $r(K)$ obtained
from the numerical
simulation of the equations
of motion (2.20). The
computation in the latter case
is explained in the text

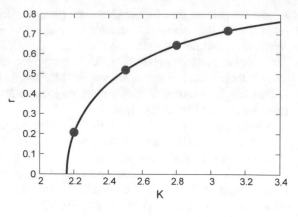

Fig. 2.5 Full lines: the
function $h(r)$, defined by the
right hand side of Eq. (1.50),
for three different values of
K (denoted close to each
curve), two below and one
above the threshold value
$K_c = 2/\pi$. The dashed line
indicates the height equal
to 1

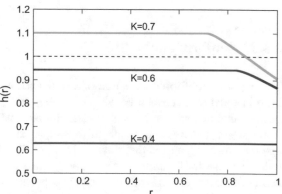

In the limit of vanishing noise, the threshold value of the coupling constant is according to Eq. (1.51) given by $K_c = 2/(\pi g(0)) = 2/\pi \approx 0.6366$ (we remind that we are taking $\omega_1 = 0.5$). The function $r(K)$ is obtained by computing from Eq. (1.50) the positive solutions for r. To this end, we plot in Fig. 2.5 the right hand side of Eq. (1.50) as a function of r for three different values of K, two below and one above the threshold value K_c. In the figure, this function is denoted by $h(r)$; it may be expressed in terms of elementary functions, but its explicit expression is not important here. In all cases, the function $h(r)$ is constant for r smaller than a K-dependent value; this range of r includes all its possible values between 0 and 1 if K is smaller than 0.5. We see that as long as we have $K < K_c$, there is no positive solution of Eq. (1.50) (as for any symmetric unimodal distribution function), since we have $h(r) < 1$ in the whole range. However, as soon as one has $K > K_c$, the function $h(r)$ intersects the horizontal line at height 1 at a finite positive value of r. It is simple to find that for $K \to K_c^+$, this value of r is equal to $\pi/4$ (incidentally, this last value is the same for any choice of ω_1, while obviously K_c depends on ω_1).

The conclusion of this analysis is the presence in the noiseless case of a first-order synchronization transition, whereby r jumps from 0 to the value $r = \pi/4$ at $K = K_c$.

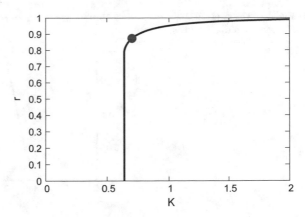

Fig. 2.6 Full line: the function $r(K)$ for a uniform frequency distribution with $\omega_1 = 0.5$ and for $D = 0$, obtained by solving the self-consistent equation (1.50). Circle: values of $r(K)$ obtained from the numerical simulation of the equations of motion (1.16) for $K = 0.7$

It is then obvious that Eq. (1.56), based on a power expansion, cannot provide a small positive value of r for K close to K_c. In Fig. 2.6, we plot the solution of Eq. (1.50) with the results of simulations for $K = 0.7$, a value slightly larger than K_c.

2.4 Nonunimodal $g(\omega)$

As soon as the frequency distribution $g(\omega)$ is no more symmetric and unimodal, the analysis becomes more difficult, and the set of possible stable states of the system gets richer. As a matter of fact, at least for noisy synchronizing systems, much less work has been done, while some results have been obtained for noiseless (i.e., $D = 0$) systems. In this section, we briefly consider the study of noisy Kuramoto systems with a particular symmetric bimodal frequency distribution, while we dismiss altogether the study of even more general $g(\omega)$. We hope that this example, for which although only some possible features will be presented here, will give a flavor of the behavior of the system for frequency distributions not belonging to the class of symmetric unimodal functions.

Let us consider a frequency distribution given by the superposition of two Gaussians of equal width, one centered at a positive frequency ω_1 and the other centered at the corresponding negative frequency $-\omega_1$:

$$g(\omega) = \frac{1}{2\sqrt{2\pi\sigma^2}}\left[e^{-\frac{(\omega-\omega_1)^2}{2\sigma^2}} + e^{-\frac{(\omega+\omega_1)^2}{2\sigma^2}}\right]. \tag{2.70}$$

The function $g(\omega)$ is bimodal if we have $\omega_1 > \sigma$; in that case, there is a minimum at $\omega = 0$ and two symmetric maxima at $\pm\omega_M$, with the value of $\omega_M < \omega_1$ depending on σ and ω_1. On the other hand, if $\omega_1 < \sigma$, there is only a single maximum at $\omega = 0$, and $g(\omega)$ falls in the class of unimodal functions.

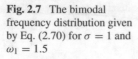

Fig. 2.7 The bimodal frequency distribution given by Eq. (2.70) for $\sigma = 1$ and $\omega_1 = 1.5$

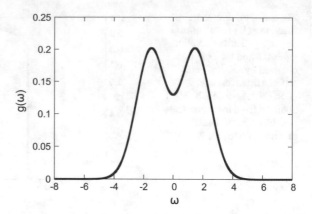

A first hint of the greater complexity resulting from a bimodal distribution comes from the fact that now we have more parameters that play a role in the behavior of the system. While for a unimodal distribution, we had K, σ and D, which actually entered into the analysis through the two quantities K/D and σ/D, we now have K, σ, ω_1 and D, and, by normalizing three of the parameters using the fourth one, we are left with three quantities. In our analysis, we take $\sigma = 1$, which is tantamount to considering the behavior of the system as a function of K/σ, ω_1/σ and D/σ. However, we also fix the value of ω_1, choosing $\omega_1 = 1.5$. Doing things this way of course does not let us explore the whole parameter space, but nevertheless will suffice to show the various different types of behavior of the system. In Fig. 2.7, we plot $g(\omega)$ for the case $\sigma = 1$, $\omega_1 = 1.5$. We are going to analyze the behavior of the system for several values of the parameter D. The critical coupling constant K_c is obtained, as before, from Eq. (2.36), while the self-consistent equation (2.34) allows to compute as a function of K, when it exists, the value of the order parameter r for the stationary solutions of the Fokker-Planck equation (we remind that the numerical solution of Eq. (2.34) can be obtained with the method described in Appendix 3).

In Fig. 2.8, we show the results obtained for five different values of D, i.e., $D = 0$, $D = 0.1$, $D = 0.5$, $D = 1$ and $D = 1.5$. The full thin line shows the function for the noiseless case, $D = 0$. The form of the curve for K close to K_c, in particular, the fact that a positive r exists for $K < K_c$, is what is expected on the basis of Eq. (2.39), since $g''(0)$ is positive. This feature is retained for small noise ($D = 0.1$, dotted line, and $D = 0.5$, dashed line) and also for the case with $D = 1$ (dot-dashed line, although in the figure, the feature is hardly visible). On the other hand, for the highest value of noise considered, $D = 1.5$ (full thick line), the function $r(K)$ has a shape similar to the unimodal case; in particular, for K close to K_c, the order parameter r is positive for $K > K_c$. We see that, except for $D = 1.5$, in the cases considered, there is a range of K to the left of K_c where there are two values of the stationary order parameter r; furthermore, the value of K_c has a nonmonotonic behavior with respect to D. We comment on these facts below.

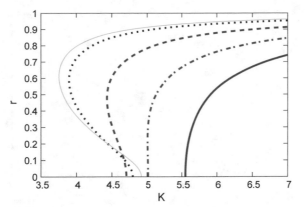

Fig. 2.8 The function $r(K)$, giving the value of the order parameter r for the stationary solutions of the Fokker-Planck equation for the bimodal frequency distribution (2.70) with $\sigma = 1$ and $\omega_1 = 1.5$, for five different values of the parameter D, namely, $D = 0$ (full thin line), $D = 0.1$ (dotted line), $D = 0.5$ (dashed line), $D = 1$ (dot-dashed line), $D = 1.5$ (full thick line). The values of the critical coupling constant are: $K_c \approx 4.915$ for $D = 0$, $K_c \approx 4.792$ for $D = 0.1$, $K_c \approx 4.701$ for $D = 0.5$, $K_c \approx 5.016$ for $D = 1$, $K_c \approx 5.548$ for $D = 1.5$. For the first four values of D, the minimum values K_m of the corresponding curves are: $K_m \approx 3.755$ for $D = 0$, $K_m \approx 3.898$ for $D = 0.1$, $K_m \approx 4.432$ for $D = 0.5$, $K_m \approx 5.007$ for $D = 1$

However, what we mention above is not the end of the story, since the situation is more complex with respect to the case with a unimodal frequency distribution also as regards the stability of the incoherent state. For the unimodal distribution, we have found that the incoherent state (we remind that it exists as a stationary state of the Fokker-Planck equation for any value of K) is unstable exactly for those values of K for which a stationary state with a positive r exists, i.e., for $K > K_c$. With a function like the full thin curve in Fig. 2.8, one has to determine if the incoherent state is stable or unstable also for K smaller than K_c. To study this problem, one has to compute the solution of the dispersion relation (2.54), or, equivalently, of the two equations (2.55) and (2.56), giving the real and the imaginary part of the dispersion relation, respectively. However, contrary to the case of a unimodal distribution, now the solution λ can be complex, so that we are not left with Eq. (2.58) and the search for its real solutions.

The incoherent state is unstable when the solution of Eq. (2.54) has a positive real part. Equation (2.55) shows that if the dispersion relation has a solution, then its real part λ_r must be larger than $-D$; then the incoherent state is stable if we have $-D < \lambda_r < 0$, and in particular, we have observed that for the noiseless case $D = 0$, this implies that the incoherent state is at most neutrally stable.

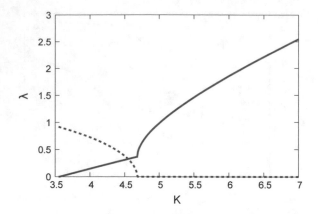

Fig. 2.9 The solution of the dispersion relation (2.54) for the bimodal frequency distribution (2.70) with $\sigma = 1$ and $\omega_1 = 1.5$, in the noiseless case $D = 0$. The full line gives the real part, while the dashed line shows the imaginary part (for values of K for which the imaginary part does not vanish, there are two complex conjugate solutions, and in the figure, we show without loss of generality the solution with positive imaginary part). For K smaller than a value ≈ 3.55, the dispersion relation has no solution, and the incoherent state is neutrally stable. The imaginary part of λ vanishes for K larger than a value ≈ 4.68; correspondingly, there is a kink in the curve for the real part

2.4.1   The Noiseless Case, $D = 0$

Let us begin the aforementioned stability analysis with the noiseless case, for which the function $r(K)$ is shown with the thin full line in Fig. 2.8. In Fig. 2.9, we plot the solution λ of the dispersion relation as a function of K. We see that for $K > K_u \approx 3.55$, the incoherent state is unstable. However, for $K < K_m \approx 3.755$ (see caption of Fig. 2.8), there is no stationary state with positive order parameter r. So, what is happening? In Fig. 2.10, we plot the results of a simulation of the noiseless system with $K = 3.7$, starting from the unstable uniform initial condition. We may observe that the system settles to what looks like a periodic nonstationary state. We thus conclude that in this range of K, i.e., for $K_u < K < K_m$, there is no stable stationary state, and the system settles asymptotically to a periodic state.

The situation is even more interesting for K slightly larger than K_m. In this case, as in the range $K_m < K < K_c$, for a given K, there are two solutions of the dispersion relation, i.e., two stationary states with a positive r. In analogy with what happens for subcritical transitions, one could guess that only one of such stationary states is stable,[8] the one with larger r. In Fig. 2.11, we plot the results of two simulations performed at $K = 3.85$; the initial state of one simulation is the uniform state, while the initial state of the other is one with large r. The plot clearly shows that there are different asymptotic states, namely, a periodic state and a stable partially coherent

---

[8] We emphasize that our analytical stability analysis concerns only the incoherent state; the stability analysis of a partially coherent state is generally a much harder task. However, if the system studied by numerical simulations settles to such a state, we may infer that it is stable.

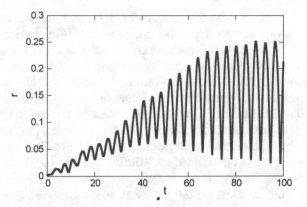

Fig. 2.10 Simulation of the system (2.20) with bimodal frequency distribution in the noiseless case $D = 0$, and with $K = 3.7$. The system, initially prepared in the unstable uniform state, settles to what seems to be a periodic state

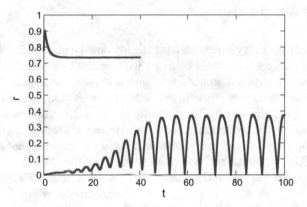

Fig. 2.11 Two simulations of the system (2.20) with bimodal frequency distribution in the noiseless case $D = 0$, and with $K = 3.85$. If the system is initially prepared in the unstable uniform state, it settles to a periodic state (lower curve); on the other hand, if the initial state has a high value of r (upper curve), the system goes to a stable stationary state with positive r. The latter simulation has been shorter, and for this reason the upper curve is plotted only up to a certain time

state. The system selects one of the two depending on its initial state. These results also confirm that the stationary state with smaller value of r is unstable. For larger values of K, although still smaller than K_c, the periodic state disappears, and only the stable partially coherent state remains. We have not performed simulations to evaluate the approximate value of K where this change occurs, but have only determined that it is smaller than the value of K where the imaginary part of λ vanishes (see Fig. 2.9 and its caption). On further increase of K, the situation does not change anymore (except that at $K = K_c$, the unstable stationary state with positive r disappears, but obviously this is not seen in simulations).

We emphasize that a complete analysis of the noiseless case with a bimodal symmetric frequency distribution is performed in Ref. [6] for the case where the distribution is a superposition of two Lorentzians, rather than two Gaussians. A technique similar to that described in Sect. 1.7.2 allows to reduce the infinite dimensional dynamical problem to a four-dimensional problem (just as for a unimodal Lorentzian, the technique allowed to reduce it to a one-dimensional dynamics). Then, the different dynamical regimes in the (ω_1, σ) parameter space are found to be separated by several types of bifurcations of the low-dimensional dynamical problem. In the same work, it is guessed that an analogous picture holds with two Gaussians, although in that case, much less can be obtained analytically. The interested reader is referred to that work for full details.[9] In our example, where we have chosen a fixed value $\sigma = 1$, we have not explored the full parameter space. However, further comments on this are given when we consider the noisy cases with $D > 0$.

2.4.2   The Noisy Cases

We now analyze the noisy cases. We have already noted that, as in Fig. 2.8, the value of K_c does not have a monotonic behavior with respect to D. However, and this is what we expect on physical grounds, the figure shows also that for any given K, the value of r of the stable partially coherent state decreases monotonically with D.

Considering first the case with small noise, i.e., $D = 0.1$, we show in Fig. 2.12 the solution λ of the dispersion relation. For K smaller than a value ≈ 3.54, the dispersion relation has no solution, and the incoherent state is stable. It remains stable up to a value $K_u \approx 3.87$, when the real part of λ becomes positive. We have found that the different dynamical regimes and the associated transitions are similar to the noiseless case. In particular, for $K_u < K < K_m \approx 3.898$ (see caption of Fig. 2.8), the system settles to a periodic state; for K larger than K_m up to a given value less than K_c, the system can settle either to the periodic state or to a stable stationary state with positive r. Increasing further the value of K, the periodic state disappears, and only the stable partially coherent state remains (as for the noiseless case, we have not determined the approximate value where this last transition occurs).

For larger noise, the situation changes. In Fig. 2.13, we plot the solution λ of the dispersion relation for $D = 0.5$ and $D = 1$. Now the solution, when it exists, is always real, and there is no need to plot the vanishing imaginary part. In both cases, the dispersion relation has no solution for K smaller than a value ≈ 4.68; in this range of values of K, the incoherent state is stable. However, it is stable up to the value of K where λ becomes positive, which for both the values of D coincides with the critical value K_c that is about 4.701 for $D = 0.5$ and about 5.016 for $D = 1$.

---

[9]Just to anticipate some information to the interested reader, we note that our transition at $K = K_u$, with the appearance of the periodic state, corresponds to a Hopf bifurcation; the transition at K_m, where a stable stationary state appears, corresponds to a saddle-node bifurcation; while the transition where the periodic state disappears corresponds to a homoclinic bifurcation.

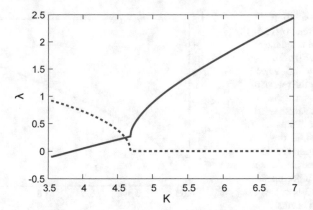

Fig. 2.12 The solution of the dispersion relation (2.54) for the bimodal frequency distribution (2.70) with $\sigma = 1$ and $\omega_1 = 1.5$, in the case with $D = 0.1$. The meaning of the full line and dashed lines is as in Fig. 2.9. For K smaller than a value ≈ 3.54, the dispersion relation has no solution, and the incoherent state is stable. The real part of λ becomes positive at a value of $K = K_u \approx 3.87$, where the incoherent state becomes unstable. The imaginary part of λ vanishes for K larger than a value equal to about 4.68; correspondingly, there is a kink in the curve for the real part

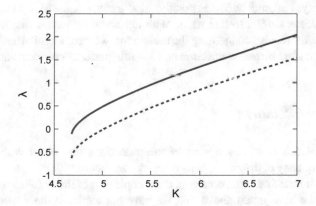

Fig. 2.13 The solution of the dispersion relation (2.54) for the bimodal frequency distribution (2.70) with $\sigma = 1$ and $\omega_1 = 1.5$, in the cases with $D = 0.5$ (full line) and $D = 1$ (dashed line). Only the real part of λ is plotted, since the imaginary part vanishes identically. For both the values of D, and for K smaller than a value ≈ 4.68, the dispersion relation has no solution, and the incoherent state is stable. It becomes unstable when λ becomes positive, which occurs $K = K_c$, i.e., at $K \approx 4.701$ for $D = 0.5$ and $K \approx 5.016$ for $D = 1$

Therefore, contrary to what happens for smaller noise, there is a K range, namely, $K_m < K < K_c$, where both the incoherent state and the partially coherent state are stable (we remind that for $K < K_m$, the partially coherent state does not exist). This range is very small for $D = 1$, when one has $K_m \approx 5.007$, but it is more extended for $D = 0.5$, when one has $K_m \approx 4.432$ (see caption of Fig. 2.8). So, now, on increasing

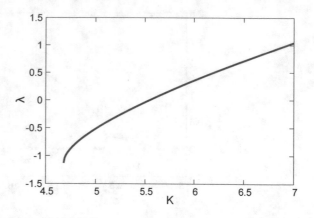

Fig. 2.14 The solution of the dispersion relation (2.54) for the bimodal frequency distribution (2.70) with $\sigma = 1$ and $\omega_1 = 1.5$, when $D = 1.5$. Only the real part of λ is plotted, since the imaginary part vanishes identically. For K smaller than a value ≈ 4.68, the dispersion relation has no solution, and the incoherent state is stable. It becomes unstable when λ becomes positive, which occurs $K = K_c$, i.e., at $K \approx 5.548$

the value of K, we have a first transition where the stable partially coherent state appears, and then a second transition where the incoherent state becomes unstable.

We finally consider the case with $D = 1.5$. The solution of the dispersion relation is plotted in Fig. 2.14. Similar to the previous two cases, the solution λ, when it exists, is always real, and becomes positive at K exactly equal to K_c. However, now the curve $r(K)$ is similar to that found with unimodal frequency distributions (see Fig. 2.8). Therefore, also concerning the transitions, we have a similarity: at $K = K_c$, the incoherent state becomes unstable and a stable partially coherent state appears.

2.4.3 Final Remarks

At the end of this brief description of the transitions that occur with a bimodal Gaussian frequency distribution, some remarks are in order.

The most interesting thing concerns the comparison of the noiseless case and the noisy case. We have noted above that we have not explored the whole parameter space, since we have a fixed variance for all the value of ω_1 that determines the centers of the two superimposed Gaussian. In Ref. [6], where the noiseless case with two superimposed Lorentzians is studied, we may consider fixed the value of the noise at $D = 0$, while the ranges of the other parameters are fully considered. As a consequence, for $D = 0$, we have found only some of the possible transition types. However, and this is the interesting feature, what we have missed at $D = 0$ reappears for large noise. We are referring in particular to the situation in which both the incoherent and the partially coherent states are stable, and to the case where the system behaves as if the frequency distribution was unimodal. For $D = 0$, this would occur by gradually increasing the width σ with respect to ω_1; instead, we have obtained this by gradually increasing the noise D at fixed σ and ω_1. In other words, as an increase of σ at fixed ω_1 "softens" the bimodal character of the distribution, an increase of D

at fixed σ and ω_1 has the same effective result.[10] This may be understood on physical grounds: the noise makes the proper frequency of each oscillator "fluctuate", the more so the larger the noise, and at sufficiently large noise, this fluctuation is comparable to the distance between the two superimposed Gaussians; this is physically equivalent to increasing σ with respect to ω_1.

Another remark concerns a technical point. The reader may have noticed that a value of $K \approx 4.68$ occurs for all cases considered. For $D = 0$ and $D = 0.1$, it is the value where the imaginary part of λ becomes zero, while for $D = 0.5$, $D = 1$ and $D = 1.5$, where the dispersion relation has only real solutions, it is the value below which such solution does not exist. This feature may be understood by looking at Eq. (2.55). For $\lambda_i = 0$, we see that in this equation, only the combination $\lambda_r + D$ appears, and therefore, if for $D = D_1$, there is a solution λ_r, then for $D = D_2$, the solution is given by $\lambda_r + D_1 - D_2$. This implies that when the solution λ is real, there is a constant difference, varying K, for the various values of D (this may be observed by comparing Figs. 2.9, 2.12, 2.13 and 2.14), and also that the appearance of a real solution occurs at the same value of K.

2.5 Beyond the Kuramoto Model

One may ask about the behavior of the system when the interaction $f(\theta)$ in the equations of motion (2.1) and the distribution of frequencies $g(\omega)$ are both generic. This more general case has received much less attention, and it is no doubt considerably more difficult to treat than the Kuramoto model. Pioneering work has been done by Daido for the noiseless case [7–10]. The extension to the case with noise has been studied by Crawford [11, 12], with the attention focussed on the scaling behavior of the order parameter near the onset of the synchronization transition. In this section, we discuss some of the results that have been obtained, without any claim of completeness. The interested reader may consult the cited bibliography for more details.

In our brief description, we are not going to consider the more general case, that is, with both $f(\theta)$ and $g(\omega)$ completely generic. Interesting new features appear even if we restrict to a unimodal symmetric frequency distribution function. Therefore, for definiteness, we will concentrate on the case of a generic interaction $f(\theta)$ (although posing a requirement, as explained shortly) with a unimodal symmetric distribution function $g(\omega)$.

The most generic interaction in the equations of motion (2.1) may be written in the form of the Fourier expansion

---

[10] Again, for the interested reader, the transition occurring at $D = 0.5$ and $D = 1$ at $K = K_m$, where the partially coherent state appears, is the analogous of a saddle-node bifurcation at $D = 0$ (as for smaller noise, but with the difference that it occurs with the presence of a stable incoherent state), while the transition where the incoherent state becomes unstable at $K = K_c$ is an example of a transcritical bifurcation.

$$f(\theta) = \sum_{k=-\infty}^{+\infty} c_k e^{ik\theta}, \qquad (2.71)$$

where k is restricted to integer values in order that $f(\theta)$ is 2π-periodic, and where convergence of the infinite series on the right hand side is obviously assumed. The coefficient c_0 may be put equal to 0 without loss of generality, by redefining ω_i, if necessary. The condition that $f(\theta)$ is real requires that we should have $c_{-k} = c_k^*$ (where we remind that the star denotes complex conjugation). However, we will impose a further requirement, namely, that c_k is purely imaginary; this is equivalent to restricting to an interaction function that contains only terms proportional to $\sin(k\theta)$ and not terms proportional to $\cos(k\theta)$. This requirement means that we want to consider only interactions that can be derived from a potential, although the analysis could be performed in the more general case. For the Kuramoto model, where $f(\theta) = -\sin\theta$, we have $c_1 = -c_{-1} = i/2$, while all the other c_k's vanish. Now, by defining $c_k = -c_{-k} = ib_k/2$, the interaction becomes

$$f(\theta) = -\sum_{k=1}^{\infty} b_k \sin\theta. \qquad (2.72)$$

We require that b_1 is positive, and although one could consider the case in which the other Fourier coefficients can be of either sign, we restrict to the case in which we have $b_k \geq 0$ for each $k > 1$; in other words, all the Fourier components of the interaction are attractive. The equations of motion then become

$$\frac{d\theta_i}{dt} = \omega_i - \frac{K}{N}\sum_{k=1}^{\infty} b_k \sum_{j=1}^{N} \sin(k\theta_i - k\theta_j) + \eta_i(t); \quad i = 1, 2, \ldots, N, \qquad (2.73)$$

with the statistical properties of the noise $\eta_i(t)$ given by Eq. (2.2).

As for the Kuramoto model, one may write down the corresponding Fokker-Planck equation corresponding to the Langevin dynamics (2.73), as

$$\frac{\partial}{\partial t}\rho(\theta, \omega, t) = -\omega\frac{\partial}{\partial\theta}\rho(\theta, \omega, t) + D\frac{\partial^2}{\partial\theta^2}\rho(\theta, \omega, t)$$

$$+ K\sum_{k=1}^{\infty} b_k \frac{\partial}{\partial\theta}\left\{\left[\int d\omega' g(\omega')\int d\theta' \sin(k\theta - k\theta')\rho(\theta', \omega', t)\right]\rho(\theta, \omega, t)\right\}.$$

$$(2.74)$$

It is possible to extend the analogy of the procedure adopted for the Kuramoto model by defining complex order parameters by

$$r_k(t)e^{ik\psi_k(t)} = \int d\omega \int_0^{2\pi} d\theta\, g(\omega)e^{ik\theta}\rho(\theta, \omega, t); \qquad k = 1, 2, \ldots, \qquad (2.75)$$

so that the Fokker-Planck equation may be rewritten as

$$\frac{\partial}{\partial t}\rho(\theta,\omega,t) = -\frac{\partial}{\partial\theta}\left[\left(\omega + K\sum_{k=1}^{\infty}b_k r_k(t)\sin(k\psi_k(t)-k\theta)\right)\rho(\theta,\omega,t)\right]$$
$$+ D\frac{\partial^2}{\partial\theta^2}\rho(\theta,\omega,t). \tag{2.76}$$

We assume the existence of a stationary solution of the above Fokker-Planck equation, to which the distribution function $\rho(\theta,\omega,t)$ tends asymptotically in time. In this stationary state, all the complex order parameters are time independent, i.e., $r_k(t)$ and $\psi_k(t)$ tend to asymptotic values r_k and ψ_k. The equation determining the stationary distribution is

$$-\frac{\partial}{\partial\theta}\left[\left(\omega + K\sum_{k=1}^{\infty}b_k r_k\sin(k\psi_k-k\theta)\right)\rho(\theta,\omega)\right] + D\frac{\partial^2}{\partial\theta^2}\rho(\theta,\omega) = 0. \tag{2.77}$$

Redefining if necessary the origin of θ, we may assume that $\psi_1 = 0$ when we have $r_1 > 0$. On physical grounds, we may assume that also all the other ψ_k's for $k > 1$ will be equal to 0 when we have the corresponding $r_k > 0$. On these premises, the time-independent Fokker-Planck equation may be rewritten as

$$-\frac{\partial}{\partial\theta}\left[\left(\omega - K\sum_{k=1}^{\infty}b_k r_k\sin(k\theta)\right)\rho(\theta,\omega)\right] + D\frac{\partial^2}{\partial\theta^2}\rho(\theta,\omega) = 0. \tag{2.78}$$

This equation generalizes Eq. (2.31) that was for the Kuramoto model. Its solution for given values of the order parameters r_k may be written in a convenient form by introducing the function

$$V(\theta) \equiv K\sum_{k=1}^{\infty}\frac{b_k}{k}r_k\cos(k\theta), \tag{2.79}$$

which is just the negative of the potential from which the interaction can be derived. Then, the 2π-periodic solution of Eq. (2.78) is given by

$$\rho(\theta,\omega;\{r_k\}) = Ce^{\frac{V(\theta)+\omega\theta}{D}}\left[1 + \left(e^{-\frac{2\pi\omega}{D}}-1\right)\frac{\int_0^{\theta}d\theta'e^{-\frac{V(\theta')+\omega\theta'}{D}}}{\int_0^{2\pi}d\theta'e^{-\frac{V(\theta')+\omega\theta'}{D}}}\right], \tag{2.80}$$

where $C = \rho(0,\omega;\{r_k\})\exp(-V(0)/D)$ is the normalization constant. This solution is acceptable when the self-consistent equations

$$r_k = \int d\omega\int_0^{2\pi}d\theta\, g(\omega)\cos(k\theta)\rho(\theta,\omega;\{r_k\}); \qquad k = 1,2,\dots \tag{2.81}$$

are satisfied. The homogeneous solution $\rho = 1/(2\pi)$ always exists, in which all the order parameter r_k's vanish, but, as for the Kuramoto case, we expect it to be unstable for sufficiently large K.

To proceed in a manner analogous to the Kuramoto case, we should now determine as a function of the coupling coefficient K when the order parameters r_k's become positive. It is not feasible to give a general answer to this question without having some additional information on the Fourier coefficients b_k. In the following, we will show how the value of the critical coupling coefficient K_c at which the synchronizing transition takes place depends on the Fourier coefficients b_k.

For K sufficiently small, we expect that the only stationary solution is the homogeneous one, and that on increasing K, there is a synchronizing transition in which the order parameters become positive.[11] When the Fourier coefficients b_k for $k > 1$ are sufficiently small with respect to b_1, we expect that the system behaves in a way not completely different from the Kuramoto model in which only b_1 is different from 0. Then, at the synchronization transition, r_1 acquires a positive value (together, in general, with all the others r_k, as argued in footnote 11). However, for general Fourier coefficients b_k, things could be different. It is not difficult to see that the critical value K_c of the coupling coefficient is given by an expression very similar to that obtained in the Kuramoto model. To this end, we may proceed as follows.

We first compute the expansion of the solution (2.80) containing only the terms that are of first order in the order parameters r_k. One obtains

$$
\rho(\theta, \omega; \{r_k\}) =
$$
$$
= C \left\{ 1 + \frac{K}{D} \sum_{k=1}^{\infty} \frac{r_k b_k}{k} \frac{\omega_2 + k^2 D^2 \cos(k\theta) + k\omega D \sin(k\theta)}{\omega^2 + k^2 D^2} \right\} + O(\{r_k^2\}), \quad (2.82)
$$

where the last term on the right hand side denotes loosely all terms of higher order.[12] Plugging this expression in the self-consistent equations (2.81), we obtain at first order the equation

$$
r_k = \frac{1}{2} K r_k k b_k \int d\omega \frac{g(\omega)}{\omega^2 + k^2 D^2} = \frac{1}{2} K r_k k b_k \int d\omega \frac{g(D\omega)}{\omega^2 + k^2}; \qquad k = 1, 2, \ldots .
$$
$$
(2.83)
$$

For any k, this expression gives a value of the coupling coefficient K, let us call it $K_c^{(k)}$, that could potentially be the critical value of the synchronizing transition; it is the value for which the factor multiplying r_k in the right hand side is equal to 1. Then, we have

---

[11] We note the following. As soon as a given r_p in the solution (2.80) is positive, we expect that in general, provided the corresponding b_k does not vanish, the right hand side of Eq. (2.81) will be different from 0 not only for $k = p$, but for all k that are multiples of p; in particular, if we have $r_1 > 0$, the right hand side of (2.81) will not vanish in general for any k.

[12] Actually, mathematical rigor would require to expand also the normalization constant C. However, since the integral of $\cos(p\theta)$ multiplied by the expression in curly brackets in Eq. (2.82) does not contain zeroth-order terms, the expansion of C is not necessary.

$$K_c^{(k)} = 2 \left[\int d\omega \, g(D\omega) \frac{kb_k}{\omega^2 + k^2} \right]^{-1}. \tag{2.84}$$

Clearly, the real critical coupling coefficient K_c will be the smallest of these values. Consider, for example, Fourier coefficients b_k such that kb_k decreases when k increases. From Eq. (2.84), we see that this condition is sufficient to have that $K_c^{(k)}$ increases with k. In this case, the critical coupling coefficient K_c is $K_c^{(1)}$, i.e., it is determined by the first Fourier coefficient b_1. We expect that at the synchronization transition, all the order parameters r_k acquire a positive value.[13] On the other hand, if, e.g., the Fourier coefficient b_2 is much larger than all the others, we expect that the critical coupling coefficient K_c is given by $K_c^{(2)}$. In that case, at the synchronization transition, we expect that r_2 acquires a positive value, with r_1 remaining equal to 0 for K sufficiently close to K_c.

We now study in analogy with the Kuramoto model the stability of the incoherent state. The procedure is very similar to that adopted for the Kuramoto model. To linearize the Fokker-Planck equation (2.74), we write

$$\rho(\theta, \omega, t) = \frac{1}{2\pi} + \delta\rho(\theta, \omega, t), \tag{2.85}$$

assuming that $\delta\rho(\theta, \omega, t) \ll 1$, which we substitute in the Fokker-Planck equation, and then keep only the terms linear in $\delta\rho$. We obtain

$$\frac{\partial}{\partial t}\delta\rho(\theta, \omega, t) = -\omega\frac{\partial}{\partial\theta}\delta\rho(\theta, \omega, t) + D\frac{\partial^2}{\partial\theta^2}\delta\rho(\theta, \omega, t)$$
$$+ \frac{K}{2\pi}\sum_{k=1}^{\infty} kb_k \int d\omega' g(\omega') \int d\theta' \cos(k\theta - k\theta')\delta\rho(\theta', \omega', t). \tag{2.86}$$

Using the Fourier expansion

$$\delta\rho(\theta, \omega, t) = \sum_{k=-\infty}^{+\infty} \widehat{\delta\rho}_k(\omega, t)e^{ik\theta}, \tag{2.87}$$

we obtain the following equation for the k-th Fourier component:

$$\frac{\partial}{\partial t}\widehat{\delta\rho}_k(\omega, t) = -ik\omega\widehat{\delta\rho}_k(\omega, t) - Dk^2\widehat{\delta\rho}_k(\omega, t)$$
$$+ \frac{K}{2}kb_k \int d\omega' g(\omega')\widehat{\delta\rho}_k(\omega', t). \tag{2.88}$$

Unlike the Kuramoto model, now there is a contribution from the interaction for any Fourier component (unless the corresponding b_k vanishes), but we have the

---

[13]This was the situation studied by Daido in the noiseless case [7, 8].

simplifying property that the equations of the different components decouple from one another. By posing

$$\widehat{\delta\rho}_k(\omega, t) = \widetilde{\delta\rho}_k(\omega, \lambda)e^{\lambda}t, \tag{2.89}$$

we obtain upon substituting in Eq. (2.88) the result

$$\left(\lambda + ik\omega + Dk^2\right)\widetilde{\delta\rho}_k(\omega, \lambda) = \frac{K}{2}kb_k \int d\omega' g(\omega')\widetilde{\delta\rho}_k(\omega', \lambda). \tag{2.90}$$

From this equation, one can obtain besides the continuous spectrum of stable modes decaying with rate Dk^2 also the discrete spectrum. To this end, we rewrite the last equation as

$$\widetilde{\delta\rho}_k(\omega, \lambda) = \frac{Kkb_k}{2\left(\lambda + ik\omega + Dk^2\right)} \int d\omega' g(\omega')\widetilde{\delta\rho}_k(\omega', \lambda). \tag{2.91}$$

Multiplying both sides by $g(\omega)$, and then integrating, we obtain the dispersion relation

$$\frac{K}{2} \int d\omega \frac{kb_k g(\omega)}{\lambda + ik\omega + Dk^2} = 1. \tag{2.92}$$

Following exactly the same argument as adopted for Eqs. (2.55), (2.56) and (2.57), we can show that the solutions λ of Eq. (2.92) can be only real. Then, the dispersion relation may be rewritten as

$$\frac{K}{2} \int d\omega g(\omega) \frac{kb_k(\lambda + Dk^2)}{\left(\lambda + Dk^2\right)^2 + k^2\omega^2} = 1. \tag{2.93}$$

Following now the same argument as was used for Eqs. (2.59) and (2.60), we may prove that Eq. (2.93) can have at most one solution, which it is necessarily larger than $-Dk^2$. Then, the instability threshold is given by the value of K for which the solution of the equation is $\lambda = 0$. We thus obtain

$$K_c^{(k)} = 2\left[\int d\omega g(D\omega)\frac{kb_k}{\omega^2 + k^2}\right]^{-1}, \tag{2.94}$$

which is same as Eq. (2.84).

For the class of unimodal symmetric frequency distribution functions $g(\omega)$, we have thus obtained a result that is very similar to the one obtained for the Kuramoto model: the instability threshold $K_c^{(k)}$ coincides with the value of K related to the onset of a positive value of the order parameter r_k. However, we must be careful! We have emphasized earlier that the synchronizing transition occurs at the smallest value of K among the $K_c^{(k)}$'s, which is denoted with K_c that contains no superscript. According to the last result on the instability thresholds, the incoherent state will be unstable for any K larger than K_c. The additional information provided by this last result is that for $K < K_c^{(k)}$, the incoherent state is stable with respect to perturbation modes with wavenumber k, but if at the same time we have $K > K_c$, there will be at least a wavenumber such that the incoherent state is unstable with respect to perturbation modes with that wavenumber.

We conclude by just citing the result related to the scaling of the order parameters for K close to K_c. The interested reader can find full details in the works of Crawford [11, 12]. We remind that in the Kuramoto model, we found that for K close to K_c, the order parameter r scales as $(K - K_c)^{1/2}$, and that this scaling behavior occurs both in the noisy case ($D > 0$) as well as in the noiseless case ($D = 0$). For the case of generic interaction, the study of the scaling behavior is more difficult. Crawford was able to analyze it through the application of the center manifold reduction [13]. To cite the result, let us consider for definiteness a case in which the instability threshold K_c is given by $K_c^{(1)}$, and the interaction has also the second Fourier component, i.e., $b_2 > 0$. Then the result is that for K sufficiently close to K_c, the scaling of the order parameter r_1 for $D > 0$ is as $(K - K_c)^{1/2}$; but this behavior does not go over to the $D = 0$ case where the scaling is as $(K - K_c)$. From the practical point of view, when the noise D is small, one has to go extremely close to K_c to observe a crossover from a scaling with exponent 1 to a scaling with exponent $1/2$. When D tends to 0, one should go infinitely close to K_c, so that the scaling remains with exponent 1 up to K_c. This behavior stems from a singularity in the cubic term of the amplitude equation arising from the center manifold reduction; the singularity arises as soon as one has $b_2 > 0$. In this sense, the scaling with exponent $1/2$ found for $D = 0$ in the Kuramoto model is non-generic.

A final remark: For the Kuramoto model with a bimodal frequency distribution $g(\omega)$, we have found that for certain ranges of the parameters (the coupling constant K, the noise intensity D, the distance between the peaks of the bimodal distribution), the distribution function $\rho(\theta, \omega, t)$ does not tend asymptotically to a stationary distribution, but rather to a distribution with periodic evolution. We expect that such cases will be present also with generic interaction functions when we relax the requirement of a unimodal frequency distribution.

Appendix 1: An H-Theorem for a Particularly Simple Case

We have seen in Sect. 2.2 that when all the oscillators have the same natural frequency ω_0, the passage to a rotating frame of reference allows to write a Fokker-Planck equation in the form that for convenience we rewrite here:

$$\frac{\partial}{\partial t}\rho(\widetilde{\theta}, \omega_0, t) = -\frac{\partial}{\partial\widetilde{\theta}}\left[Kr(t)\sin(\psi(t) - \widetilde{\theta} - \omega_0 t)\rho(\widetilde{\theta}, \omega_0, t)\right] + D\frac{\partial^2}{\partial\widetilde{\theta}^2}\rho(\widetilde{\theta}, \omega_0, t).$$

$$(2.95)$$

The corresponding Langevin equations are

$$\frac{d\widetilde{\theta}_i}{dt} = -\frac{K}{N} \sum_{j=1}^{N} \sin(\widetilde{\theta}_i - \widetilde{\theta}_j) + \eta_i(t), \qquad i = 1, 2, \ldots, N. \qquad (2.96)$$

We note that the above equations are the noisy version of the dynamics of a system of oscillators that interact according to the interparticle potential energy $U = -(K/(2N)) \sum_{i,j} \cos(\widetilde{\theta}_i - \widetilde{\theta}_j)$. We may express the energy of the system at time t using the one-body distribution function $\rho(\widetilde{\theta}, \omega_0, t)$, by using the same approximation as was employed in the derivation of the Fokker-Planck equation for ρ, i.e., neglect of correlations between the oscillators. We may simply obtain (without writing from now on the unnecessary ω_0 dependence in order not to overload the notation[14]

$$u(t) \equiv \frac{U(t)}{N} = -\frac{K}{2} \left[\left(\int_0^{2\pi} d\widetilde{\theta} \, \cos\widetilde{\theta} \, \rho(\widetilde{\theta}, t) \right)^2 + \left(\int_0^{2\pi} d\widetilde{\theta} \, \sin\widetilde{\theta} \, \rho(\widetilde{\theta}, t) \right)^2 \right]$$

$$= -\frac{K}{2} \left[r_x^2(t) + r_y^2(t) \right], \qquad (2.97)$$

where in the second line, we have indicated the two components of the order parameter in the rotating frame. With the time evolution of $\rho(\widetilde{\theta}, t)$ determined by Eq. (2.95), we may study the evolution of the free energy per particle $f(t) = u(t) - Ds(t)$, where $s(t)$ is the entropy per particle [2] given by[15]

$$s(t) = -\int_0^{2\pi} d\widetilde{\theta} \, \rho(\widetilde{\theta}, t) \ln \rho(\widetilde{\theta}, t). \qquad (2.98)$$

In the rotating frame, the phase of the order parameter is $\widetilde{\psi}(t) = \psi(t) - \omega_0 t$, so that we may write down the Fokker-Planck equation (2.95) as

$$\frac{\partial}{\partial t} \rho(\widetilde{\theta}, t) = \frac{\partial}{\partial \widetilde{\theta}} \left[K \left(r_x(t) \sin\widetilde{\theta} - r_y(t) \cos\widetilde{\theta} \right) \rho(\widetilde{\theta}, t) \right] + D \frac{\partial^2}{\partial \widetilde{\theta}^2} \rho(\widetilde{\theta}, t). \qquad (2.99)$$

Using this equation and the expressions (2.97) and (2.98), it is straightforward to obtain

$$\frac{df}{dt} = -\int_0^{2\pi} d\widetilde{\theta} \left[K r_x(t) \sin\widetilde{\theta} - K r_y(t) \cos\widetilde{\theta} + \frac{D}{\rho(\widetilde{\theta}, t)} \frac{\partial \rho(\widetilde{\theta}, t)}{\partial \widetilde{\theta}} \right]^2 \rho(\widetilde{\theta}, t) \leq 0. \qquad (2.100)$$

Thus, $f(t)$ can only decrease. Moreover, since it may be easily seen that $f(t)$ is bounded from below, the evolution will stop when the expression in the square brackets in the integral on the right hand side of the last equation vanishes. We

---

[14]Although unnecessary, we maintain the tilde also when $\widetilde{\theta}$ is an integration variable, just as a reminder that we are considering the dynamics in the rotating frame of reference.

[15]The parameter D plays the role of the temperature.

therefore conclude that the normalized distribution function ρ will tend to (we remind that the Fokker-Planck equation conserves the normalization)

$$\rho(\widetilde{\theta}) = Ae^{\frac{K}{D}[r_x \cos \widetilde{\theta} + r_y \sin \widetilde{\theta}]}, \tag{2.101}$$

obtained by setting the expression in the square brackets in Eq. (2.100) to zero. Here, A is the normalization constant, while r_x and r_y have to satisfy self-consistent equations. Without loss of generality, we may choose the origin of the angle $\widetilde{\theta}$ such that $r_y = 0$; we then obtain the expression (2.26) for $\rho(\widetilde{\theta})$, with $r_x = r$ satisfying Eq. (2.28).

The derivation of this H-theorem is based on the possibility to write an expression of a potential energy u for the system. For this particular case with only one frequency ω_0, this may be done in the rotating frame of reference, where the equations of motion may be written in the form (2.96). In the general case with distributed natural frequencies, the frequencies will appear in any frame of reference (except for those oscillators with a frequency equal to that of the rotating frame). This prevents the possibility to have a potential energy in the form $U = -(K/(2N)) \sum_{i,j} \cos(\widetilde{\theta}_i - \widetilde{\theta}_j)$, and there will be a contribution $\sim \omega_i \theta_i$ for those oscillators that have a nonzero frequency in the rotating frame. In fact, the expression $\omega\theta$ is not an acceptable energy, since it is not 2π-periodic. As a consequence, the approach of the distribution $\rho(\theta, \omega, t)$ to a particular asymptotic state cannot be proved in the manner done above. Nevertheless, in the analyses presented in this chapter, it is assumed that such an approach occurs anyways (this may be directly verified in simulations).

Appendix 2: Form of the Function $r(K)$ for Symmetric and Unimodal Frequency Distributions in the Kuramoto Model

The function $r(K)$ is obtained by computing for each K the solution for r of the self-consistent equation (2.34). For concreteness, let us refer to the right hand side of this equation as $M(r)$ when seen as a function of r. We know the derivative at $r = 0$ of the right hand side of equation (2.34): it is given by the coefficient of r in the first term on the right hand side of Eq. (2.35) (either the first or the second line). This positive derivative increases proportionally to K, and K_c is the value for which it is equal to 1. We also know on physical grounds (but it may also be proved analytically) that $M(1)$ tends to 1 as $K \to \infty$. If we could prove that for any K, the function $M(r)$ has a negative second derivative for any $r > 0$, this would show that there is for each $K > K_c$ only one solution of Eq. (2.34), and that this solution $r(K)$ increases with K, tending to 1 as $K \to K_c$. Such a behavior would be similar to what happens for the noiseless case, where we could prove the aforementioned properties. If these properties hold, then, for any symmetric unimodal distribution $g(\omega)$, $r(K)$ would have the behavior indicated by the full line in Fig. 2.1, as we have found in

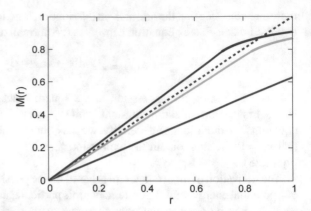

simulations for the Gaussian, the Lorentzian and the uniform distribution. It is not difficult to see that a negative definite second derivative of $M(r)$ is sufficient for our purpose. First, the derivative of $M(r)$ is positive for $r = 0$ and for r sufficiently large (since $M(r)$ must tend asymptotically to 1 from below). Then, a negative definite second derivative implies that the first derivative is always positive. Second, since $M(r)$ depends on r only through the product Kr, a negative definite second derivative and a positive definite first derivative imply that the equality $M(r) = r$, i.e., the self-consistent Eq. (2.34), can be satisfied for only one value of r such that $0 < r < 1$ (for $K > K_c$), and that this value increases with K (the last property arising from the fact that if for a particular value $K = K_1$, we have $M(r) = r$ for $r = r_1$, then for $K > K_1$, it will be $M(r_1) > r_1$, so that $M(r) = r$ will be satisfied for a value of r larger than r_1).

Now, we give the argument in favor of the fact that the second derivative of $M(r)$ is negative definite. By defining with $M(r, \omega)$ the function obtained by integrating the right hand side Eq. (2.34) with respect to θ only, we may write

$$M(r) = \int d\omega \, g(\omega) M(r, \omega). \qquad (2.102)$$

Let us consider what happens for a uniform distribution equal to $1/(2\omega_1)$ between $\omega = -\omega_1$ and $\omega = \omega_1$. When ω_1 becomes very large, we expect that the presence of the noise with a finite coefficient D has very little influence, since most of the proper frequencies are very large, and their contribution to the equations of motion is much stronger than that of the noise. Practically, we expect that $M(r)$ will behave for sufficiently large ω_1 as if one has $D = 0$. But we know what is $M(r)$ for $D = 0$: referring to the function $h(r)$ plotted in Fig. 2.5, it will be $M(r) = rh(r)$. This may be inferred from the fact that the self-consistent equation (1.50) was obtained by dividing by r both members of Eq. (1.49), since in the latter equation, r was a common factor in both the members. Then we obtain the plots in Fig. 2.15.

Here we have not denoted the values of K for each curve, since it depends on the concrete value of ω_1. However, we have found that for $K < \omega_1$ (that was equal to 0.5 in Sect. 2.3.3), we are in the situation of the lowest of the three curves in Fig. 2.15, which is a straight line, thus with vanishing second derivative. But when ω_1 increases without limits, any finite K will be smaller than ω_1. We conclude that if we pose $g(\omega)$ equal to a constant in Eq. (2.102), $M(r)$ is a straight line.

Now, here comes the crucial assumption, which we could verify numerically, but for which we have no proof. We assume that for any given r, $M(r, \omega)$ has a negative second r-derivative for $|\omega| < \omega^*(r)$ and a positive second r-derivative for $|\omega| > \omega^*(r)$ (we know that for $\omega = 0$, the second r-derivative is negative, since $M(r, 0)$ is the ratio of the modified Bessel functions of argument (Kr/D) of first and zeroth order, respectively). Then, for a symmetric unimodal $g(\omega)$, we may write

$$
\begin{aligned}
\frac{d^2 M(r)}{dr^2} &= \int d\omega\, g(\omega) \frac{\partial^2 M(r, \omega)}{\partial r^2} \\
&< \int_{|\omega|<\omega^*(r)} d\omega\, g(\omega^*(r)) \frac{\partial^2 M(r, \omega)}{\partial r^2} + \int_{|\omega|>\omega^*(r)} d\omega\, g(\omega^*(r)) \frac{\partial^2 M(r, \omega)}{\partial r^2} \\
&= g(\omega^*(r)) \int d\omega\, \frac{\partial^2 M(r, \omega)}{\partial r^2} = g(\omega^*(r)) \times 0 = 0;
\end{aligned}
\tag{2.103}
$$

This inequality is what we wanted to obtain.

Appendix 3: The Numerical Solution of Eq. (2.34)

The numerical solution of Eq. (2.34) would involve the discretization of the two integrations with respect to θ and ω. In order to avoid the integration in θ, we have preferred to employ a Fourier expansion in θ, since this allows a faster computation. Let us then start from the time independent Fokker-Planck equation (2.31) that we write in the equivalent form given by Eq. (2.41). The equation is rewritten here for convenience:

$$
(\omega - Kr \sin\theta)\, \rho(\theta, \omega; r) - D \frac{\partial}{\partial\theta} \rho(\theta, \omega; r) = S(\omega),
\tag{2.104}
$$

where $S(\omega)$ is the constant and uniform probability current in the stationary state, and $\rho(\theta, \omega; r)$ is the normalized stationary distribution function given in Eq. (2.32). However, we would not need the last explicit expression, since we will only use the expression

$$
\int_0^{2\pi} d\theta\, e^{ip\theta} \rho(\theta, \omega; r) = \langle e^{ip\theta} \rangle(\omega; r)
\tag{2.105}
$$

of the expectation value (which depends on ω and r) of the function $e^{ip\theta}$ for the distribution $\rho(\theta, \omega; r)$, where p is any positive integer.

Multiplying Eq. (2.104) by $e^{ip\theta}$ and integrating over θ, we obtain for any given ω the following system of equations:

$$(\omega + iD) \langle e^{i\theta} \rangle(\omega; r) + i\frac{Kr}{2} \langle e^{i2\theta} \rangle(\omega; r) = i\frac{Kr}{2} \tag{2.106}$$

$$(\omega + ipD) \langle e^{ip\theta} \rangle(\omega; r) + i\frac{Kr}{2} \langle e^{i(p+1)\theta} \rangle(\omega; r)$$

$$-i\frac{Kr}{2} \langle e^{i(p-1)\theta} \rangle(\omega; r) = 0, \qquad p = 2, 3, \ldots \tag{2.107}$$

Solving this system, one obtains $\langle e^{ip\theta}(\omega; r) \rangle$ for each p. Then, the self-consistent equation (2.34) may be written as

$$r = \int d\omega \, g(\omega) \mathrm{Re} \left[\langle e^{i\theta}(\omega; r) \rangle \right]. \tag{2.108}$$

Therefore, we first solve the systems of Eqs. (2.106)–(2.107); then we use the solution in Eq. (2.108), thus performing only one numerical integration (in ω). This procedure is considerably faster than numerical integration with a discrete integration step; however, like the latter, it also introduces an approximation, i.e., the truncation of the system at a convenient value of p. In our computations, we have truncated the system at $p = 50$; we have checked that higher truncation did not change in any appreciable way the value of the quantity $e^{i\theta}(\omega; r)$, which is the one used in Eq. (2.108) to compute r.

References

1. H. Risken, *The Fokker-Planck Equation: Methods of Solution and Applications* (Springer, Berlin, 1996)
2. K. Huang, *Statistical Mechanics* (Wiley, New York, 1987)
3. H. Sakaguchi, Prog. Theor. Phys. **79**, 39 (1988)
4. S.H. Strogatz, Physica D **143**, 1 (2000)
5. S.H. Strogatz, *Nonlinear Dynamics And Chaos: With Applications To Physics, Biology, Chemistry, And Engineering* (Westview Press, Boulder, 2014)
6. E.A. Martens, E. Barreto, S.H. Strogatz, E. Ott, P. So, T.M. Antonsen, Phys. Rev. E **79**, 026204 (2009)
7. H. Daido, Progr. Theor. Phys. **88**, 1213 (1992)
8. H. Daido, Progr. Theor. Phys. **89**, 929 (1993)
9. H. Daido, Phys. Rev. Lett. **73**, 760 (1994)
10. H. Daido, Physica D **91**, 24 (1996)
11. J.D. Crawford, Phys. Rev. Lett. **74**, 4341 (1995)
12. J.D. Crawford, K.T.R. Davies, Physica D **125**, 1 (1999)
13. J. Guckenheimer, P. Holmes, *Nonlinear Oscillations, Dynamical Systems, and Bifurcations of Vector Fields* (Springer, New York, 1986)

Chapter 3
Oscillators with Second-Order Dynamics

Abstract In the first section, we introduce the generalized Kuramoto model with inertia and noise, and discuss in turn its connection to electrical power distribution networks, its interpretation as a long-range interacting system driven out of equilibrium and its dynamics written in a dimensionless form convenient for further analysis. In section two, we discuss our recent numerical results on very interesting nonequilibrium phase transitions exhibited by the model in the stationary state for the case of unimodal frequency distributions: the system shows a phase transition between a synchronized and an incoherent stationary state. Section three is devoted to an analytical treatment of the observed phase transitions, in which we discuss both the incoherent stationary state and its linear stability and the synchronized stationary state of the dynamics. In section four, we take up the issue of comparing and interpreting simulation results for a finite system vis-à-vis our derived analytical results in the thermodynamic limit, thereby providing interesting insights into the (slow) relaxation properties of the dynamics.

Keywords Oscillators with inertia and noise · Second-order dynamics
Nonequilibrium phase transitions · Kramers equation
Incoherent stationary state and linear stability · Synchronized stationary state
Slow relaxation

Until now we have studied synchronizing systems constituted by interacting limit-cycle oscillators that have a first-order dynamics in time. We have seen in Chap. 1 for the case of an isolated oscillator how introduction of inertia leads to a second-order dynamics in time, thus resulting in significant differences in the dynamical properties with respect to the first-order dynamics.

In this chapter, we focus on a rather interesting and relevant generalization of the Kuramoto dynamics (1.16) that includes stochastic noise, as considered in Chap. 2, but, more significantly, inertial terms parametrized by a moment of inertia [1–10]. Inclusion of inertia elevates the first-order Kuramoto dynamics to one that is second order in time, while as discussed in Chap. 1, noise effectively accounts for stochastic fluctuations of the dynamical parameters in time. The generalized model reduces in specific limits to the Kuramoto model with and without noise, and furthermore has the merit of offering the possibility to explore the issue of emergence of spontaneous synchronization in a wider space of parameters than that of the first-order dynamics. Moreover, the generalized model gives even with a unimodal natural frequency distribution with a non-compact support a rather rich phase diagram that includes both equilibrium and nonequilibrium phase transitions. We may remind the reader that with such a frequency distribution, the first-order model shows only a nonequilibrium second-order phase transition, as discussed in Chap. 1.

Besides proliferation in behavior with the inclusion of inertia, the generalized model offers a rather remarkable bridge between two apparently disconnected research areas, namely, the area of spontaneous synchronization pursued by dynamical physicists and that of statistical physics studies, in both in and out of equilibrium regimes, of so-called long-range interacting systems pursued within the community of statistical physicists. Indeed, it turns out that two different limits of the generalized model have been studied extensively over the years, albeit with not much overlap and inter-community dialogue, by the communities of dynamical and statistical physicists. As we will demonstrate in this chapter, the dynamics of the generalized Kuramoto model describes a long-range interacting system of particles moving on a unit circle under the influence of a set of external drive in the form of quenched external torques acting on the individual particles and in presence of stochastic noise. With the noise, but without the external torques, the resulting model is the so-called Brownian mean-field (BMF) model [11], introduced as a generalization of the celebrated Hamiltonian mean-field (HMF) model that serves as a prototype to study statics and dynamics of long-range interacting systems [12].

In recent years, there has been a surge in interest in studies of systems with long-range interactions. In these systems, the inter-particle potential in d dimensions decays at large separation r as $r^{-\alpha}$, with $0 \leq \alpha \leq d$ [13–17]. Examples are gravitational systems, plasmas, two-dimensional hydrodynamics, charged and dipolar systems, etc. Unlike systems with short-range interactions, long-range interacting systems are generically non-additive, implying that dividing the system into macroscopic subsystems and summing over their thermodynamic variables such as energy do not yield the corresponding variables of the whole system. Non-additivity leads to many significant thermodynamic and dynamical consequences, such as negative microcanonical specific heat, inequivalence of statistical ensembles, and others, which are unusual with short-range interactions [15].

In addition to the aforementioned merits of the generalized model, the latter finds its relevance in addressing a problem of practical interest. As mentioned in Chap. 1, an early motivation behind studying analytically the phenomenon of synchronization, for which the Kuramoto model serves as the prototype, was to explain the spectacular phenomenon of spontaneous synchronization among fireflies. Namely, the phenomenon observed in parts of south-east Asia of thousands of male fireflies gathering in trees at night and flashing on and off in unison. It was revealed in a study due to Ermentrout that among fireflies of a particular species (the *Pteroptyx mallacae*), the approach to synchronization from an initially unsynchronized state is faster in the Kuramoto setting than in reality [18]. Ermentrout proposed a route to reconciliation that involves accounting for finite inertia of the Kuramoto oscillators, which elevates the first-order dynamics of the Kuramoto model to the level of second-order dynamics.[1] Including also a Gaussian noise term that accounts for the stochastic fluctuations of the natural frequencies in time [19], one arrives at the generalized Kuramoto model including inertia and noise. We will discuss below a proof that the resulting dynamics leads to a nonequilibrium stationary state (NESS) at long times.

Study of NESSs is a very active area of research of modern day statistical mechanics [20]. These states are characterized by a violation of detailed balance, thereby leading to a net non-zero probability current around a closed loop in the configuration space. A primary challenge in this field is to formulate a tractable framework to analyze nonequilibrium systems on a common footing, similar to the one due to Gibbs and Boltzmann that exists for equilibrium systems.

3.1 Generalized Kuramoto Model with Inertia and Noise

The generalized Kuramoto dynamics constitutes the dynamical variable of angular velocity v_j assigned to each oscillator in addition to its phase θ_j. The equations of motion are given by [2, 3]

$$\frac{d\theta_j}{dt} = v_j, \quad m\frac{dv_j}{dt} = -\gamma v_j + \gamma \omega_j - \widetilde{K} r \sin(\theta_j - \psi) + \widetilde{\eta}_j(t), \qquad (3.1)$$

where m is the common moment of inertia of the oscillators, $\gamma > 0$ is a parameter that plays the role of a damping constant, $\widetilde{K}$ is the strength of coupling between the oscillators, while $\widetilde{\eta}_j(t)$ is a Gaussian, white noise satisfying

$$\langle \widetilde{\eta}_j(t) \rangle = 0, \quad \langle \widetilde{\eta}_j(t)\widetilde{\eta}_k(t') \rangle = 2\widetilde{D}\delta_{jk}\delta(t - t'). \qquad (3.2)$$

Here, $\widetilde{D} \geq 0$ is a parameter that sets the strength of the noise. The quantity r in Eq. (3.1) is the usual Kuramoto order parameter defined in Eq. (1.40) of Chap. 1. Note that $re^{i\psi}$ being a complex number, see Eq. (1.40), we may resolve it along the real (or the x-) axis and the imaginary (or the y-) axis, thus obtaining the quantities

$$r_x(t) \equiv r(t) \cos \psi(t) = \frac{1}{N} \sum_{j=1}^{N} \cos(\theta_j(t)),$$

$$r_y(t) \equiv r(t) \sin \psi(t) = \frac{1}{N} \sum_{j=1}^{N} \sin(\theta_j(t)),$$

$$r(t) = \sqrt{r_x^2(t) + r_y^2(t)}, \quad \psi(t) = \tan^{-1}(r_y(t)/r_x(t)). \qquad (3.3)$$

---

[1]That inertia may have significant effect on relaxation properties was already seen in Chap. 1 for the case of a single oscillator.

We consider the frequency distribution $g(\omega)$ to be unimodal with a non-compact support and symmetric about the mean $\langle \omega \rangle = 0$. We will denote the width of the distribution by σ. Note that any non-zero value of $\langle \omega \rangle$ can be trivially gotten rid off by viewing the dynamics in a frame rotating uniformly with frequency $\langle \omega \rangle$ with respect to the laboratory frame.

In the limit of overdamped motion ($m \to 0$ at a fixed $\gamma \neq 0$), the dynamics (3.1) reduces to

$$\gamma \frac{d\theta_j}{dt} = \gamma \omega_j - \widetilde{K} r \sin(\theta_j - \psi) + \widetilde{\eta}_j(t). \tag{3.4}$$

Then, defining $K \equiv \widetilde{K}/\gamma$ and $\eta_j(t) \equiv \widetilde{\eta}_j(t)/\gamma$ so that $D = \widetilde{D}/\gamma^2$, the dynamics (3.4) for $D = 0$ becomes that of the Kuramoto model, Eq. (1.16), and for $D \neq 0$ that of its noisy version given by the dynamics (2.20).

3.1.1 Relation to Electrical Power Distribution Networks

It turns out that the dynamics (3.1) without the noise term, studied in [1], also arises in the context of electrical power distribution networks comprising synchronous generators (representing power plants) and motors (representing customers) [21, 22]; the dynamics arises in the approximation in which every node of the network is connected to every other. We discuss this connection to power networks in the present subsection.

The basic elements of a power distribution network or grid are synchronous generators located at power plants and motors located with the consumers. A generator converts mechanical (or other forms of energy, e.g., nuclear energy) into electrical energy, while the reverse is true for a motor. Let P denote the power, which being generated is a positive quantity for a generator and being consumed is negative for a motor. Either a generator or a motor basically consists of a rotating turbine whose state for the j-th unit is represented by its phase

$$\theta_j(t) = \Omega t + \phi_j(t), \tag{3.5}$$

where Ω is the standard supply frequency, $\Omega = 50/60$ Hz typically, while $\phi_j(t)$ is the deviation from uniform rotation. From considerations of energy conservation, we easily see that the generated or consumed power P_i^{source} of the i-th element has to equal the sum of the power P_i^{trans} exchanged with the grid, the power $P_i^{\text{acc}} = (I/2)(d/dt)(d\theta_i(t)/dt)^2$ accumulated in the turbine, and the amount $P_i^{\text{diss}} = \kappa(d\theta_i(t)/dt)^2$ dissipated in overcoming friction. Here, I is the moment of inertia of the turbine, while κ is the friction constant. The power transmitted between two elements j and i that are connected by a transmission line would depend on the phase difference across the ends of the transmission line, and may be shown to be given by $P_{\text{max};ji} \sin(\theta_j - \theta_i)$, where $P_{\text{max};ji}$ is the maximum capacity of the transmission line. With $P_i^{\text{trans}} = \sum_j P_{\text{max};ji} \sin(\theta_j - \theta_i)$, we then arrive at the equation

$$P_i^{\text{source}} = \frac{I}{2} \frac{\mathrm{d}}{\mathrm{d}t} \left(\frac{\mathrm{d}\theta_i(t)}{\mathrm{d}t} \right)^2 + \kappa \left(\frac{\mathrm{d}\theta_i(t)}{\mathrm{d}t} \right)^2 - \sum_j P_{\text{max};ji} \sin(\theta_j - \theta_i). \qquad (3.6)$$

With the assumption that $|\mathrm{d}\phi/\mathrm{d}t| \ll \Omega$, one then obtains the dynamics [21, 22]

$$\frac{\mathrm{d}^2 \phi_i(t)}{\mathrm{d}t^2} = P_i - \gamma \frac{\mathrm{d}\phi_i}{\mathrm{d}t} + \sum_j K_{ji} \sin(\phi_j - \phi_i), \qquad (3.7)$$

where we have

$$P_i = \frac{P_i^{\text{source}} - \kappa \Omega^2}{I \Omega}, \qquad (3.8)$$

$$\gamma = \frac{2\kappa}{I}, \qquad (3.9)$$

$$K_{ji} = \frac{P_{\text{max};ji}}{I \Omega}. \qquad (3.10)$$

In the mean-field approximation in which every unit i is connected to every other unit j with equal strength and $K_{ji} = K/N$, where N is the total number of nodes in the network, Eq. (3.7) may be reduced to the form

$$\frac{\mathrm{d}^2 \phi_i(t)}{\mathrm{d}t^2} = P_i - \gamma \frac{\mathrm{d}\phi_i}{\mathrm{d}t} + \frac{K}{N} \sum_j \sin(\phi_j - \phi_i). \qquad (3.11)$$

Let us note that the P_i's are intrinsic to the units and would vary in general from one unit to another, so that they may be regarded as quenched random variables. The above dynamics is similar to the generalized Kuramoto model dynamics (3.1) in the absence of noise $\eta_i(t)$, which may be additionally considered also in the dynamics (3.11) as accounting for random ambient noise.

3.1.2 The Model as a Long-Range Interacting System

We now show that in a different context than that of coupled oscillators, the dynamics (3.1) describes a long-range interacting system of particles moving on a unit circle, with each particle acted upon by a quenched external torque $\widetilde{\omega}_j \equiv \gamma \omega_j$. Recent exploration of long-range interacting systems, and in particular, of their static and dynamic properties, has focussed on an analytically tractable and representative model called the Hamiltonian mean-field (HMF) model [12].

The HMF model consists of N particles of mass m that are moving on a unit circle and are interacting through a long-range interparticle potential that is of the mean-field type: every particle is coupled to every other with equal strength. The Hamiltonian of the HMF model is given by [12]

$$H = \sum_{j=1}^{N} \frac{p_j^2}{2m} + \frac{\widetilde{K}}{2N} \sum_{j,k=1}^{N} \left[1 - \cos(\theta_j - \theta_k)\right], \qquad (3.12)$$

where $\theta_j \in [-\pi, \pi]$ is the position of the j-th particle on the circle, while $p_j = m v_j$ is its conjugated angular momentum, with v_j being the angular velocity. The time evolution of the system within a microcanonical ensemble takes place following the deterministic Hamilton equations of motion:

$$\frac{\mathrm{d}\theta_j}{\mathrm{d}t} = v_j, \quad m \frac{\mathrm{d}v_j}{\mathrm{d}t} = -\widetilde{K} r \sin(\theta_j - \psi). \qquad (3.13)$$

The dynamics evidently conserves the total energy and momentum, and moreover leads at long times to an equilibrium stationary state in which, depending on the energy density $\epsilon \equiv H/N$, the system could be in one of two possible phases. Namely, for ϵ smaller than a critical value $\epsilon_c = 3\widetilde{K}/4$, the system is in a clustered phase in which the particles are close together on the circle, while for $\epsilon > \epsilon_c$, the particles are uniformly distributed on the circle, thus characterizing a homogeneous phase [15]. A continuous phase transition between the two phases is characterized by a positive value of stationary-state r, denoted by r_{st}, in the clustered phase and a zero value in the homogeneous phase.

It is natural to invoke a generalization of the microcanonical dynamics (3.13) to account for interaction with an external heat bath at temperature T. The resulting model, which goes by the name of the Brownian mean-field (BMF) model, has thus a canonical-ensemble dynamics given by [11].

$$\frac{\mathrm{d}\theta_j}{\mathrm{d}t} = v_j, \quad m \frac{\mathrm{d}v_j}{\mathrm{d}t} = -\gamma v_j - \widetilde{K} r \sin(\theta_j - \psi) + \widetilde{\eta}_j(t), \qquad (3.14)$$

where $\widetilde{\eta}_j(t)$ is as in Eq. (3.2). We may employ the fluctuation-dissipation relation to express the strength $\widetilde{D}$ of the noise in terms of the temperature T and the damping constant γ as $\widetilde{D} = \gamma k_B T$ [20]. We will set the Boltzmann constant k_B to unity in the rest of the chapter for reasons of convenience. The canonical dynamics (3.14) also leads to a long-time equilibrium stationary state in which a generic configuration $C \equiv \{\theta_j, v_j\}_{1 \le j \le N}$ that has energy $E(C)$ occurs with the Gibbs-Boltzmann weight: $P_{\mathrm{eq}}(C) \propto \exp[-E(C)/T]$. The phase transition in the HMF model observed within the microcanonical ensemble now occurs within the canonical ensemble as one tunes the temperature across the critical value $T_c = \widetilde{K}/2$. The derivation of this result is discussed in Sect. 3.3.1.

We now consider a set of quenched external torques $\{\widetilde{\omega}_j \equiv \gamma \omega_j\}$ acting on each of the particles, thereby pumping energy into the system. In this case, the second equation in the canonical dynamics (3.14) is augmented by the term $\widetilde{\omega}_j$ on the right hand side. The resulting dynamics becomes exactly the same as the dynamics (3.1) of the generalized Kuramoto model.

3.1.3 Dynamics in a Reduced Parameter Space

We will find it convenient to reduce the number of parameters in the dynamics given by Eq. (3.1). To achieve this, note that the effect of σ may be made explicit by replacing ω_j in the second equation by $\sigma\omega_j$. We will thus from now on consider the dynamics (3.1) with the substitution $\omega_j \to \sigma\omega_j$. In the resulting model, $g(\omega)$ is then to be regarded as having zero mean and unit width. Moreover, we will consider in the dynamics (3.1) the parameter $\widetilde{D}$ to be $\widetilde{D} = \gamma T$, see our discussions above.

For $m \neq 0$, in terms of dimensionless quantities [5, 6]

$$\bar{t} \equiv t\sqrt{\widetilde{K}/m}, \ \bar{v}_j \equiv v_j\sqrt{m/\widetilde{K}}, \ 1/\sqrt{\bar{m}} \equiv \gamma/\sqrt{\widetilde{K}m}, \ \bar{\sigma} \equiv \gamma\sigma/\widetilde{K},$$
$$\bar{T} \equiv T/\widetilde{K}, \ \bar{\eta}_j(\bar{t}) \equiv \tilde{\eta}_j(t)/\widetilde{K}, \tag{3.15}$$

the equations of motion (3.1) take the form

$$\frac{d\theta_j}{d\bar{t}} = \bar{v}_j, \quad \frac{d\bar{v}_j}{d\bar{t}} = -\frac{1}{\sqrt{\bar{m}}}\bar{v}_j - r\sin(\theta_j - \psi) + \bar{\sigma}\omega_j + \bar{\eta}_j(\bar{t}), \tag{3.16}$$

where we have

$$\langle \bar{\eta}_j(\bar{t})\bar{\eta}_k(\bar{t}')\rangle = 2\frac{\bar{T}}{\sqrt{\bar{m}}}\delta_{jk}\delta(\bar{t} - \bar{t}'). \tag{3.17}$$

For $m = 0$, defining the dimensionless time $\bar{t} \equiv t(\widetilde{K}/\gamma)$, with $\bar{\sigma}$ and $\bar{T}$ as defined above, the dynamics has the form of an overdamped motion:

$$\frac{d\theta_j}{d\bar{t}} = \bar{\sigma}\omega_j - r\sin(\theta_j - \psi) + \bar{\eta}_j(\bar{t}), \tag{3.18}$$

with $\langle \bar{\eta}_j(\bar{t})\bar{\eta}_k(\bar{t}')\rangle = 2\bar{T}\delta_{jk}\delta(\bar{t} - \bar{t}')$. We thus consider from now on instead of the dynamics (3.1) involving five parameters given by $m, \gamma, \widetilde{K}, \sigma, T$ the reduced dynamics (3.16) (or (3.18) in the overdamped limit) that involves three dimensionless parameters given by $\bar{m}, \bar{T}, \bar{\sigma}$. We will drop the overbars over various symbols for brevity of notation. With $\sigma = 0$ (i.e. $g(\omega) = \delta(\omega)$) the dynamics (3.16) becomes that of the BMF model with an equilibrium stationary state. For other choice of $g(\omega)$, we show in Appendix 1 that the dynamics (3.16) violates detailed balance leading to a NESS [5].

3.2 Nonequilibrium First-Order Synchronization Phase Transition: Simulation Results

Here, we report simulation results on a very interesting nonequilibrium phase transition that occurs in the stationary state of the dynamics (3.16). In the three-dimensional space spanned by the parameters (m, T, σ), we first locate the phase transitions in the Kuramoto model, Eq. (1.15), and in its noisy extension, Eq. (2.20).

The phase transition of the Kuramoto dynamics ($m = T = 0, \sigma \neq 0$) corresponds to a continuous transition from a low-σ synchronized to a high-σ incoherent phase occurring across the critical point given by

$$\sigma_c(m = 0, T = 0) = \frac{\pi g(0)}{2}, \tag{3.19}$$

which is obtained using Eq. (1.51), see Ref. [6]. Extending the Kuramoto dynamics to $T \neq 0$ (the noisy Kuramoto model studied in Chap. 2), the critical point (3.19) becomes a second-order critical line on the (T, σ)-plane, which is obtained by solving the equation

$$2 = \int_{-\infty}^{\infty} d\omega \, \frac{g(\omega)T}{T^2 + \omega^2 \sigma_c^2 (m = 0, T))}, \tag{3.20}$$

as may be seen by rewriting appropriately Eq. (2.36) [6]. The transition in the BMF dynamics ($m, T \neq 0, \sigma = 0$) corresponds now to a continuous transition at the critical temperature $T_c = 1/2$, a result proved in Sect. 3.3.1.

Figure 3.1a shows schematically the complete phase diagram of the model (3.16), in which the red second-order critical lines denote the continuous transitions mentioned above [5, 6]. For general non-zero values of m, σ, T, however, the synchronization transition becomes first order, occurring across the shaded blue transition surface shown in the figure. The surface is bounded by the second-order critical lines on the (T, σ) and (m, T) planes, and also by a first-order transition line on the (m, σ)-plane. We remind the reader that all phase transitions for $\sigma \neq 0$ are in NESSs.

That the phase transition for general non-zero values of m, T, σ is first order becomes evident on analyzing results of N-body simulations of the dynamics (3.16) for a representative $g(\omega)$, for example, a Gaussian distribution $g(\omega) = \exp(-\omega^2/2)/\sqrt{2\pi}$ [5, 6]. For details on the simulation procedure, we refer the reader to Appendix 2. For given values of m and T, and an initial state with all the oscillators at $\theta = 0$ and with their angular velocities sampled from a Gaussian distribution with zero mean and standard deviation $\propto T$, we first allowed the system to equilibrate at $\sigma = 0$. The state was subsequently allowed to evolve under the condition of σ increasing adiabatically to high values and back in a cycle. Figure 3.2a shows the behavior of r for several values of m at a fixed value of T smaller than the BMF transition point $T_c = 1/2$. In the figure, one may observe sharp jumps and hysteresis behavior that are hallmarks of a first-order transition. With the decrease of m, the jumps in

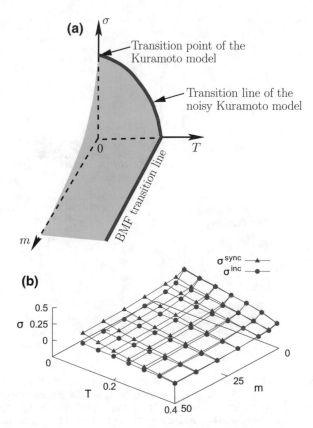

Fig. 3.1 Panel (**a**) shows the schematic phase diagram of the model (3.16) in the three-dimensional space spanned by the parameters, the dimensionless moment of inertia m, the temperature T, and the width of the frequency distribution σ. In the figure, the shaded blue surface denotes a first-order transition surface, while the thick red lines are second-order critical lines. The surface and the lines are such that the system has a synchronized stationary state inside the region bounded by the surface, and has instead an incoherent stationary state in the region outside the surface. In the figure, we also show the transitions of the noiseless and the noisy Kuramoto model and the BMF model. The blue surface in (**a**) is bounded from above and below by the dynamical stability thresholds $\sigma^{sync}(m, T)$ and $\sigma^{inc}(m, T)$ that correspond respectively to the synchronized and the incoherent phase. These thresholds are estimated in N-body simulations, in particular, from hysteresis plots of the type shown in Fig. 3.2. The surfaces $\sigma^{sync}(m, T)$ and $\sigma^{inc}(m, T)$ obtained in numerical simulations for $N = 500$ and with a Gaussian $g(\omega)$ with zero mean and unit width are displayed in panel (**b**). https://doi.org/10.1088/1742-5468/14/08/R08001 ©*SISSA Medialab Srl. Reproduced by permission of IOP Publishing. All rights reserved.*

r become less sharp, and the hysteresis loop area decreases, both features lending credence to the fact that the transition becomes second-order-like as $m \to 0$, see Fig. 3.1a. For $m = 1000$, we show in Fig. 3.2a the approximate stability thresholds for the incoherent and the synchronized state, denoted respectively by $\sigma^{inc}(m, T)$ and $\sigma^{sync}(m, T)$. The actual phase transition point $\sigma_c(m, T)$ lies in between the two

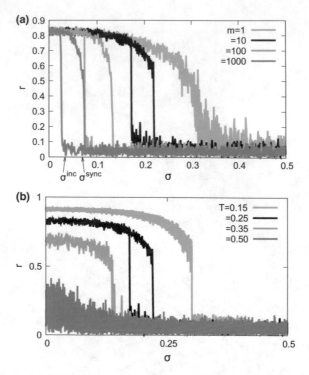

Fig. 3.2 For the model (3.16), the figure shows in (**a**) r as a function of adiabatically-tuned σ for different values of m at $T = 0.2 < T_c = 1/2$ (with T_c being the BMF transition point), and also the stability thresholds, $\sigma^{\text{inc}}(m, T)$ and $\sigma^{\text{sync}}(m, T)$, for $m = 1000$, and in (**b**) r as a function of adiabatically tuned σ for different temperatures $T \leq T_c = 1/2$ at a fixed moment of inertia $m = 10$. For a given m in (**a**), the branch of the plot to the right (left) corresponds to σ increasing (decreasing); for $m = 1$, note that the two branches almost overlap. For a given T in (**b**), the branch of the plot to the right (left) corresponds to σ increasing (decreasing); for $T \geq 0.45$, the two branches almost fall on top of one another. The data are obtained from numerical integration of the dynamics (3.16) for $N = 500$ and for a Gaussian $g(\omega)$ with zero mean and unit width. https://doi.org/10.1088/1742-5468/14/08/R08001

thresholds. We note from the figure that both the thresholds decrease and approach zero with the increase of m. Figure 3.2b shows hysteresis plots for a Gaussian $g(\omega)$ at a fixed m and for several values of $T \leq T_c$. One may observe from the figure that with T approaching T_c, the hysteresis loop area decreases, jumps in r become less sharp, and moreover, the jumps occur between smaller and smaller values that approach zero. Moreover, the r value at $\sigma = 0$ decreases as T approaches T_c, reaching zero at T_c. From these findings, we conclude that the thresholds $\sigma^{\text{inc}}(m, T)$ and $\sigma^{\text{sync}}(m, T)$ coincide on the second-order critical lines, as expected, and moreover, the lines come asymptotically close together and approach zero as $m \to \infty$ at a fixed T. For given values of m and T and σ satisfying $\sigma^{\text{inc}}(m, T) < \sigma < \sigma^{\text{sync}}(m, T)$, we show in Fig. 3.3a the dependence of r on time in the stationary state. The figure evidently shows a bistable behavior, in which the system switches back and forth between

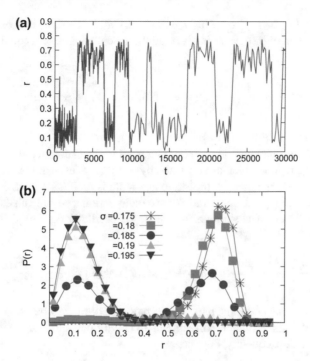

Fig. 3.3 For the dynamics (3.16) at $m = 20$, $T = 0.25$, $N = 100$, and for a Gaussian $g(\omega)$ with zero mean and unit width, panel (**a**) shows at $\sigma = 0.185$, which is the numerically estimated first-order phase transition point, the quantity r versus time in the stationary state. On the other hand, panel (**b**) shows the distribution $P(r)$ at several σ's around 0.185. The data are obtained from numerical integration of the dynamical equations (3.16) with $N = 100$. https://doi.org/10.1088/1742-5468/14/08/R08001 ©*SISSA Medialab Srl. Reproduced by permission of IOP Publishing. All rights reserved.*

incoherent ($r \approx 0$) and synchronized ($r > 0$) states. Concomitantly, the distribution $P(r)$ shown in Fig. 3.3b is bimodal with a peak around either $r \approx 0$ or $r > 0$ as σ varies between $\sigma^{\text{inc}}(m, T)$ and $\sigma^{\text{sync}}(m, T)$. Figure 3.3 is consistent with the phase transition being first order [23].

3.3 Analysis in the Continuum Limit: The Kramers Equation

In this section, we discuss analytical characterization of the dynamics (3.16) in the continuum limit $N \to \infty$. Similar to what was done for both the noiseless and the noisy Kuramoto model in Chaps. 1 and 2, let us define a single-oscillator density $f(\theta, v, \omega, t)$ that gives at time t and for each ω the fraction of oscillators that have angle θ and angular velocity v. The density f is 2π-periodic in θ, and obeys the normalization $\int_{-\pi}^{\pi} d\theta \int_{-\infty}^{+\infty} dv\, f(\theta, v, \omega, t) = 1 \,\forall\, \omega, t$. We show in Appendix 3 that the time evolution of $f(\theta, v, t)$ is given by the so-called Kramers equation [3, 5, 6, 24]

$$\frac{\partial f}{\partial t} = -v\frac{\partial f}{\partial \theta} + \frac{\partial}{\partial v}\left(\frac{v}{\sqrt{m}} - \sigma\omega + r\sin(\theta - \psi)\right)f + \frac{T}{\sqrt{m}}\frac{\partial^2 f}{\partial v^2}, \qquad (3.21)$$

with $r(t)e^{i\psi(t)} = \int d\theta dv d\omega\, g(\omega)e^{i\theta}f(\theta, v, \omega, t)$.

As in previous chapters, we will be interested in the stationary state solutions of the Kramers equation, which are obtained by setting the left hand side of Eq. (3.21) to zero. As already mentioned, the stationary state is a NESS, unless $\sigma = 0$. In the stationary state, the quantities r and ψ have their stationary-state values given by r_{st} and ψ_{st}, respectively. The stationary-state single-oscillator density $f_{st}(\theta, v, \omega)$ thus satisfies

$$0 = -v\frac{\partial f_{st}}{\partial \theta} + \frac{\partial}{\partial v}\left(\frac{v}{\sqrt{m}} - \sigma\omega + r_{st}\sin(\theta - \psi_{st})\right)f_{st} + \frac{T}{\sqrt{m}}\frac{\partial^2 f_{st}}{\partial v^2}. \qquad (3.22)$$

Similar to what was done in Chap. 1, we may set ψ_{st} to zero by choosing suitably the origin of the phase, which corresponds to having the stationary values $r_{y,st} = 0$ and $r_{x,st} = r_{st}$. Consequently, one has

$$r_{st} = \int d\theta \, dv d\omega\, g(\omega) \cos\theta f_{st}(\theta, v, \omega). \qquad (3.23)$$

From now on, we will consider the stationary-state Kramers equation with $\psi_{st} = 0$.

3.3.1 $\sigma = 0$: *Stationary Solutions and the Associated Phase Transition*

For $\sigma = 0$, the stationary-state single-oscillator density is given by the Gibbs-Boltzmann measure corresponding to canonical equilibrium [6]:

$$f_{st}(\theta, v) = \frac{\exp[-v^2/(2T) + (r_{st}/T)\cos\theta]}{\sqrt{2\pi T}\int_{-\pi}^{\pi} d\theta \,\exp[(r_{st}/T)\cos\theta]}, \qquad (3.24)$$

where the denominator ensures that one has $\int_{-\infty}^{\infty} dv \int_{-\pi}^{\pi} d\theta\, f_{st}(\theta, v) = 1$. It may be easily checked by direct substitution that the above form[2] satisfies Eq. (3.22) with $\sigma = 0$ and with $\psi_{st} = 0$. Using Eqs. (3.23) and (3.24), we arrive at

---

[2]Note that with $\sigma = 0$, all the oscillators have the same natural frequency equal to $\langle\omega\rangle$, and consequently, the need to group the oscillators based on their natural frequencies, as was done while defining the density $f(\theta, v, \omega, t)$, is no longer there. As a result, one has the stationary-state single-

$$r_{st} = \int d\theta dv \; \cos\theta f_{st}(\theta, v) = \frac{\int_{-\pi}^{\pi} d\theta \; \cos\theta \exp[(r_{st}/T)\cos\theta]}{\int_{-\pi}^{\pi} d\theta \; \exp[(r_{st}/T)\cos\theta]}. \tag{3.25}$$

The self-consistency Eq. (3.25) has a trivial solution $r_{st} = 0$ valid at all temperatures, while it may be shown that a non-zero solution exists for T smaller than a critical value $T_c = 1/2$ [15]. Reverting to dimensional temperatures on using Eq. (3.15), we obtain the critical temperature of the BMF model as $T_c = \widetilde{K}/2$, as we had promised to show towards the end of Sect. 3.1.2.

3.3.2 $\sigma \neq 0$: Incoherent Stationary State and Its Linear Stability

For $\sigma \neq 0$, the θ-independent solution characterizing the incoherent phase, for which $r_{st} = 0$, is given by [3]:

$$f_{st}^{inc}(\theta, v, \omega) = \frac{1}{2\pi} \sqrt{\frac{1}{2\pi T}} \exp\left[-\frac{(v - \sigma\omega\sqrt{m})^2}{2T}\right], \tag{3.26}$$

which may be seen by direct substitution into Eq. (3.22).

As is usual, the linear stability analysis of the incoherent state (3.26) is carried out by expanding $f(\theta, v, \omega, t)$ as $f(\theta, v, \omega, t) = f_{st}^{inc}(\theta, v, \omega) + e^{\lambda t}\delta f(\theta, v, \omega)$, with $|\delta f| \ll 1$, substituting in Eq. (3.21) with $\psi = \psi_{st} = 0$ (since we are interested in studying fluctuations in the stationary state), and then keeping terms to linear order in δf. The linearized Kramers equation is given by

$$\lambda\delta f + v\frac{\partial\delta f}{\partial\theta} - \frac{\partial}{\partial v}\left(\frac{v}{\sqrt{m}} - \sigma\omega\right)\delta f - \frac{T}{\sqrt{m}}\frac{\partial^2\delta f}{\partial v^2}$$
$$= -\frac{\partial f_{st}^{inc}}{\partial v}\int_{-\pi}^{\pi}\int_{-\infty}^{\infty}\int_{-\infty}^{\infty} d\phi dv d\omega \; g(\omega)\delta f(\phi, v, \omega)\sin(\phi - \theta). \tag{3.27}$$

Since f and f_{st}^{inc} are normalized, we have

$$\int_{-\pi}^{\pi}\int_{-\infty}^{\infty} d\theta dv \; \delta f(\theta, v, \omega) = 0. \tag{3.28}$$

On substituting

$$\delta f(\theta, v, \omega) = \sum_{n=-\infty}^{\infty} b_n(v, \omega, \lambda)e^{in\theta} \tag{3.29}$$

---

oscillator density $f_{st}(\theta, v)$ defined as the fraction of oscillators that have angle θ and angular velocity v in the stationary state.

in Eq. (3.27), we get

$$
\frac{d^2 b_n}{dv^2} + \frac{1}{T}\left(v - \sigma\omega\sqrt{m}\right)\frac{db_n}{dv} + \frac{1}{T}\left(1 - \lambda\sqrt{m} - inv\sqrt{m}\right)b_n
$$
$$
= \frac{\sqrt{m}}{T}\frac{\partial f_{st}^{inc}}{\partial v}\pi(i\delta_{n,1} - i\delta_{n,-1})\langle 1, b_n\rangle, \tag{3.30}
$$

where we have the scalar product

$$
\langle\varphi, \psi\rangle \equiv \int_{-\infty}^{\infty}\int_{-\infty}^{\infty} dv d\omega\, g(\omega)\varphi^*(v, \omega)\psi(v, \omega), \tag{3.31}
$$

with $*$ denoting complex conjugation. Since δf is real, we have $b_{-n} = b_n^*$, while Eq. (3.28) gives $b_0 = 0$. We can then focus our attention on only $n \geq 0$. Next, Eq. (3.30) may be transformed into a nonhomogeneous parabolic cylinder equation by the transformations

$$
b_n(v, \omega, \lambda) = \exp\left[-\frac{(v - \sigma\omega\sqrt{m})^2}{4T}\right]\beta_n(z, \omega, \lambda), \tag{3.32}
$$
$$
z = \frac{1}{\sqrt{T}}(v - \sigma\omega\sqrt{m} + i2nT\sqrt{m}), \tag{3.33}
$$

which when used in Eq. (3.30) yields

$$
\frac{d^2\beta_n}{dz^2} + \left[\frac{1}{2} - \frac{z^2}{4} - \sqrt{m}(\lambda + in\sigma\omega\sqrt{m} + n^2 T\sqrt{m})\right]\beta_n
$$
$$
= i\pi\sqrt{m}\frac{\partial f_{st}^{inc}}{\partial v}e^{\frac{1}{4}(z - 2i\sqrt{mT})^2}\langle 1, e^{-\frac{1}{4}(z - 2i\sqrt{mT})^2}\beta_1\rangle\delta_{n,1}. \tag{3.34}
$$

For $n \neq 1$, the right hand side of the above equation is zero, thereby giving the eigenvalues

$$
\lambda_{p,n}(\omega) = -\frac{p}{\sqrt{m}} - n^2 T\sqrt{m} - in\sigma\omega\sqrt{m}, \quad p = 0, 1, 2, \ldots, \tag{3.35}
$$

and the corresponding eigenfunctions

$$
\beta_{p,n}(z, \omega, \lambda_{p,n}) = D_p(z) = 2^{-\frac{p}{2}}e^{-\frac{z^2}{4}}H_p\left(\frac{z}{\sqrt{2}}\right), \tag{3.36}
$$

which do not depend on n and ω. Here, the quantities $D_p(z)$ and $H_p(x)$ are respectively the parabolic cylinder function and the Hermite polynomial of degree p [25]. The eigenvalues $\lambda_{p,n}(\omega)$ form a continuous spectrum. They all have negative real parts, thus leading to linear stability of the incoherent state (3.26), for $n = 2, 3, \ldots$ and $p = 0, 1, 2, \ldots$. For $n = 0$, the eigenvalues have also negative real parts unless we consider those with $p = 0$, which have a vanishing real part. They would correspond to neutrally stable modes; however, we see that the modes with $n = 0$ have zero amplitude due to the normalization condition (3.28).

For $n = 1$, solving Eq. (3.30), we get

$$\beta_1(z, \omega, \lambda) = -i\pi \langle 1, e^{-\left(\frac{z}{2} - i\sqrt{mT}\right)^2} \beta_1 \rangle$$
$$\times \sum_{p=0}^{\infty} \frac{\int_{-\infty}^{\infty} dz_1 \, e^{\left(\frac{z_1}{2} - i\sqrt{mT}\right)^2} D_p[f_{st}^{inc}]'}{\sqrt{2\pi} \, p! \left(\frac{p}{\sqrt{m}} + \lambda + i\sigma\omega\sqrt{m} + T\sqrt{m}\right)} D_p(z), \quad (3.37)$$

where

$$[f_{st}^{inc}(v)]' = \left. \frac{\partial f_{st}^{inc}}{\partial v} \right|_{v = \sigma\omega\sqrt{m} - i2T\sqrt{m} + \sqrt{T}z} = -\frac{(z - 2i\sqrt{mT})}{(2\pi)^{\frac{3}{2}} T} e^{-\frac{1}{2}(z - 2i\sqrt{mT})^2}, \quad (3.38)$$

Using the last expression to compute $\langle 1, e^{-\left(\frac{z}{2} - i\sqrt{mT}\right)^2} \beta_1 \rangle$, we obtain from the resulting self-consistent equation an eigenvalue equation for λ given by [3]:

$$\frac{2T}{e^{mT}} = \sum_{p=0}^{\infty} \frac{(-mT)^p (1 + \frac{p}{mT})}{p!} \int_{-\infty}^{\infty} \frac{g(\omega) d\omega}{1 + \frac{p}{mT} + i\frac{\sigma\omega}{T} + \frac{\lambda}{T\sqrt{m}}}. \quad (3.39)$$

In Appendix 4, we prove that the above equation has one and only one solution for λ with a positive real part, and when this single solution exists, it is necessarily real [5, 6]. A positive (respectively, negative) λ implies that the incoherent state (3.26) is linearly unstable (respectively, stable). We may then conclude that at the point of neutral stability, we have $\lambda = 0$, which substituted in Eq. (3.39) gives $\sigma^{inc}(m, T)$, the stability threshold of the incoherent stationary state, to be given by the following equation:

$$\frac{2T}{e^{mT}} = \sum_{p=0}^{\infty} \frac{(-mT)^p (1 + \frac{p}{mT})^2}{p!} \int_{-\infty}^{\infty} \frac{g(\omega) d\omega}{(1 + \frac{p}{mT})^2 + \frac{(\sigma^{inc})^2 \omega^2}{T^2}}. \quad (3.40)$$

In the (m, T, σ) space, the above equation defines the stability surface $\sigma^{inc}(m, T)$. There will also be the stability surface $\sigma^{sync}(m, T)$ for the stability threshold of the synchronized stationary state. We refer the reader to Fig. 3.1b that shows the two surfaces obtained in N-body simulations for $N = 500$ for a Gaussian $g(\omega)$.

The two surfaces, $\sigma^{inc}(m, T)$ and $\sigma^{sync}(m, T)$, coincide on the critical lines on the (T, σ) and (m, T) planes where the transition becomes continuous. On the other hand, outside these planes, the surfaces enclose the first-order transition surface $\sigma_c(m, T)$, that is, $\sigma^{sync}(m, T) > \sigma_c(m, T) > \sigma^{inc}(m, T)$, see Fig. 3.1a. Let us now show by taking suitable limits that the surface $\sigma^{inc}(m, T)$ meets the critical lines on the (T, σ) and (m, T) planes. Along the way, we will also obtain the intersection of this surface with the (m, σ)-plane. On considering $m \to 0$ at a fixed T, and on noting that only the $p = 0$ term in the sum in Eq. (3.40) contributes, we get $\lim_{m \to 0, T \text{ fixed}} \sigma^{\text{inc}}(m, T) = \sigma_c(m = 0, T)$, with the implicit expression of $\sigma_c(m = 0, T)$ given by Eq. (3.20). One also finds that $\lim_{T \to T_c^-, m \text{ fixed}} \sigma^{\text{inc}}(m, T) = 0$, that is, on the (m, T) plane, the transition line is given by $T_c = 1/2$. As $T \to 0$ at a fixed m, we get $\sigma^{\text{inc}}_{\text{noiseless}}(m) \equiv \lim_{T \to 0, m \text{ fixed}} \sigma^{\text{inc}}(m, T)$, with [5, 6].

$$1 = \frac{\pi g(0)}{2\sigma^{\text{inc}}_{\text{noiseless}}} - \frac{m}{2} \int_{-\infty}^{\infty} d\omega \, \frac{g(\omega)}{\left[1 + m^2 (\sigma^{\text{inc}}_{\text{noiseless}})^2 \omega^2\right]} . \tag{3.41}$$

3.3.3 $\sigma \neq 0$: Synchronized Stationary State

For $\sigma \neq 0$, the existence of the synchronized stationary state is borne out by our simulation results displayed in Figs. 3.2 and 3.3. For general σ, we may expand the single-oscillator density for the synchronized stationary state as [9]

$$f_{\text{st}}^{\text{sync}}(\theta, v, \omega) = \Phi_0 \left(\frac{v}{\sqrt{2T}}\right) \sum_{n=0}^{\infty} b_n(\theta, \omega) \Phi_n \left(\frac{v}{\sqrt{2T}}\right) . \tag{3.42}$$

Here, the functions b_n satisfy $b_n(\theta, \omega) = b_n(\theta + 2\pi, \omega)$ in order that $f_{\text{st}}^{\text{sync}}$ is 2π-periodic in θ. On the other hand, $\Phi_n(ax)$ is the Hermite function: $\Phi_n(ax) = \sqrt{a/(2^n n! \sqrt{\pi})} \exp(-a^2 x^2/2) H_n(ax)$, with $H_n(x)$'s being the n-th degree Hermite polynomial. The functions Φ_n are orthonormal: $\int dx \, \Phi_m(ax) \Phi_n(ax) = \delta_{mn}$. Normalization of $f_{\text{st}}^{\text{sync}}(\theta, v, \omega)$ gives $\int_{-\pi}^{\pi} d\theta \, b_0(\theta, \omega) = 1$, while the self-consistent values of the parameters r_{st} are given by

$$r_{\text{st}} = \int d\omega \, g(\omega) \int_{-\pi}^{\pi} d\theta \, b_0(\theta, \omega) \cos \theta . \tag{3.43}$$

Furthermore, using $\int dx \, x \Phi_0(ax) \Phi_n(ax) = 1/(\sqrt{2}a) \delta_{n,1}$, we obtain the result that $\int dv \, v f_{\text{st}}^{\text{sync}}(\theta, v, \omega) = \sqrt{T} b_1(\theta, \omega)$. Integrating the stationary-state Kramers equation (3.22) over v, we obtain that $\int dv \, v f_{\text{st}}^{\text{sync}}(\theta, v, \omega)$ and, hence, that $b_1(\theta, \omega)$ does not depend on θ. In choosing the Hermite functions in the expansion (3.42), we are motivated by the fact that for $\sigma = 0$, the density $f_{\text{st}}^{\text{sync}}(\theta, v, \omega)$ has the Gibbs-Boltzmann form, $f_{\text{st}}^{\text{sync}}(\theta, v, \omega) \sim \exp[-v^2/(2T) + r_{\text{st}} \cos \theta]$, cf. Equation (3.24). As may be easily seen [9], the expansion coefficients b_n for this case satisfy $b_0(\theta, 0) \sim \exp[r_{\text{st}} \cos \theta]$, $b_n(\theta, 0) = 0$ for $n > 0$, so that only the $n = 0$ term in the expansion (3.42) has to be taken into account. Then, with $\Phi_0(x) \sim \exp(-x^2/2)$, the product $\Phi_0 \left(v/\sqrt{2T}\right) \Phi_0 \left(v/\sqrt{2T}\right)$ appearing in the expansion correctly gives the velocity-part of the density $\sim \exp[-v^2/(2T)]$.

On plugging Eq. (3.42) into the stationary-state Kramers equation (3.22), on using the known recursion relations for the Hermite polynomials, and on equating to zero the coefficient of each Φ_n, we get [9]

$$\sqrt{nT}\frac{\partial b_{n-1}(\theta,\omega)}{\partial \theta} + \sqrt{(n+1)T}\frac{\partial b_{n+1}(\theta,\omega)}{\partial \theta}$$

$$+ \frac{n}{\sqrt{m}}b_n(\theta,\omega) + \sqrt{\frac{n}{T}}b_{n-1}(\theta,\omega)[r_{st}\sin\theta - \sigma\omega] = 0 \qquad (3.44)$$

for $n = 0, 1, 2, \ldots$ (with the understanding that $b_{-1}(\theta,\omega) \equiv 0$). The equation for $n = 0$ allows to recover the result that $b_1(\theta,\omega)$ is independent of θ. Noting the scaling of the various terms in Eq. (3.44) with m, we may expand $b_n(\theta,\omega)$ as [9]

$$b_n(\theta,\omega) = \sum_{k=0}^{\infty}(\sqrt{m})^k c_{n,k}(\theta,\omega), \qquad (3.45)$$

which may be shown to be an asymptotic expansion in $\sqrt{m}$ [9], thus requiring a proper numerical evaluation of the sum on the right hand side by invoking the so-called Borel summation method [26], see Appendix 6. Now, using Eq. (3.45), we conclude that $b_1(\theta,\omega)$ being independent of θ implies that so will be $c_{1,k}(\theta,\omega) \; \forall \, k$. The only constraint on $b_0(\theta,\omega)$ being $\int_{-\pi}^{\pi} d\theta \, b_0(\theta,\omega) = 1$, we may without losing generality choose $c_{0,k\geq1}(0,\omega) = 0$. We now use Eq. (3.45) in Eq. (3.44) and equate the coefficient of each power of $\sqrt{m}$ to zero. The term proportional to $\left(\sqrt{m}\right)^{-1}$ gives $nc_{n,0}(\theta,\omega) = 0$, implying $c_{n,0}(\theta,\omega) = 0$ for $n > 0$. The coefficient of the term proportional to $\left(\sqrt{m}\right)^k$ yields [9]

$$\sqrt{nT}\frac{\partial c_{n-1,k}(\theta,\omega)}{\partial \theta} + \sqrt{(n+1)T}\frac{\partial c_{n+1,k}(\theta,\omega)}{\partial \theta}$$

$$+ \sqrt{nT}a(\theta,\omega)c_{n-1,k}(\theta,\omega) + nc_{n,k+1}(\theta,\omega) = 0 \qquad (3.46)$$

for $n, k = 0, 1, 2, \ldots$ (with $c_{-1,k}(\theta,\omega) \equiv 0$), where $a(\theta,\omega) \equiv [r_{st}\sin\theta - \sigma\omega]/T$. The system of equations (3.46) can be solved recursively. The details of solving these equations may be found in Appendix 5, while we quote here only the nonvanishing solutions:

$$c_{0,0}(\theta,\omega) = c_{0,0}(0,\omega)e^{-h(\theta,\omega)}\left[1 + \left(e^{h(2\pi,\omega)} - 1\right)\frac{\int_0^\theta d\theta' e^{h(\theta',\omega)}}{\int_{-\pi}^{\pi} d\theta' e^{h(\theta',\omega)}}\right], \qquad (3.47)$$

$$c_{1,1}(\omega) = \sqrt{T}\frac{c_{0,0}(0,\omega)\left(1 - e^{h(2\pi,\omega)}\right)}{\int_{-\pi}^{\pi} d\theta' e^{h(\theta',\omega)}}, \qquad (3.48)$$

$$c_{n,n}(\theta,\omega) = -\sqrt{\frac{T}{n}}\left[\frac{\partial c_{n-1,n-1}(\theta,\omega)}{\partial \theta} + a(\theta,\omega)c_{n-1,n-1}(\theta,\omega)\right], \qquad (3.49)$$

$$c_{0,2k}(\theta,\omega) = \sqrt{2}\frac{\int_{-\pi}^{\pi} d\theta' \frac{\partial c_{2,2k}(\theta',\omega)}{\partial \theta'}e^{h(\theta',\omega)}}{\int_{-\pi}^{\pi} d\theta' e^{h(\theta',\omega)}}e^{-h(\theta,\omega)}\int_0^\theta d\theta' e^{h(\theta',\omega)}$$

$$-\sqrt{2}e^{-h(\theta,\omega)}\int_0^\theta d\theta' \frac{\partial c_{2,2k}(\theta',\omega)}{\partial \theta'} e^{h(\theta',\omega)}, \tag{3.50}$$

$$c_{1,1+2k}(\omega) = -\sqrt{2T}\frac{\int_{-\pi}^{\pi} d\theta' \frac{\partial c_{2,2k}(\theta',\omega)}{\partial \theta'} e^{h(\theta',\omega)}}{\int_{-\pi}^{\pi} d\theta' e^{h(\theta',\omega)}}, \tag{3.51}$$

$$c_{2,2+2k}(\theta,\omega) = -\sqrt{\frac{T}{2}}a(\theta,\omega)c_{1,1+2k}(\omega) - \frac{\sqrt{3T}}{2}\frac{\partial c_{3,1+2k}(\theta,\omega)}{\partial \theta}, \tag{3.52}$$

$$c_{n,n+2k}(\theta,\omega) = -\sqrt{\frac{T}{n}}\left[\frac{\partial c_{n-1,n-1+2k}(\theta)}{\partial \theta} + a(\theta,\omega)c_{n-1,n-1+2k}(\theta,\omega)\right]$$
$$-\frac{\sqrt{(n+1)T}}{n}\frac{\partial c_{n+1,n-1+2k}(\theta,\omega)}{\partial \theta} \quad n \geq 3, \tag{3.53}$$

with $k = 1, 2, \ldots$. Here, we have defined $h(\theta,\omega) \equiv \int_0^\theta d\theta' a(\theta',\omega)$.

Figure 3.11 (Appendix 5) shows in a schematic manner the flow of the solution up to $n = k = 6$, while that for higher values proceeds in an analogous manner. As shown in the figure, the system (3.46) computes progressively each element of the main diagonal, followed by the elements of the second upper diagonal, each one being determined by the knowledge of two previously determined elements, and so on. Each element of the matrix is proportional to $c_{0,0}(0,\omega)$, the latter being fixed by the normalization of f_{st}^{sync}: $\sum_{k=0}^{\infty}\int_{-\pi}^{\pi} d\theta \, (\sqrt{m})^{2k} c_{0,2k}(\theta,\omega) = 1$. The values of r_{st} have to be determined self-consistently by using Eqs. (3.43) and (3.45).

In order to illustrate an application of the aforementioned scheme, let us choose a representative $g(\omega)$, namely, a Gaussian: $g(\omega) = 1/(\sqrt{2\pi})\exp(-\omega^2/2)$, and obtain in the synchronized phase the marginal θ-distribution, defined as $n(\theta) \equiv \int_{-\infty}^{\infty} d\omega \, g(\omega)\int_{-\infty}^{\infty} dv \, f_{st}^{sync}(\theta,v,\omega)$. We also obtain a quantity proportional to the local pressure [27], given by $p(\theta) \equiv \int_{-\infty}^{\infty} d\omega \, g(\omega)\int_{-\infty}^{\infty} dv \, v^2 f_{st}^{sync}(\theta,v,\omega)$. Orthonormality of the Hermite functions implies that

$$n(\theta) = \int_{-\infty}^{\infty} d\omega \, g(\omega)b_0(\theta,\omega), \tag{3.54}$$

$$p(\theta) = T\int_{-\infty}^{\infty} d\omega \, g(\omega)\left(\sqrt{2}b_2(\theta,\omega) + b_0(\theta,\omega)\right). \tag{3.55}$$

To evaluate $n(\theta)$ and $p(\theta)$, we would thus need the coefficients $b_0(\theta,\omega)$ and $b_2(\theta,\omega)$, whose evaluation requires truncating the expansion (3.45) at suitable values k_{trunc} of k. Figure 3.11 (Appendix 5) implies that knowing $c_{2,2k}$ allows to compute $c_{0,2k}$, and so it is natural to choose the same k_{trunc} for both $b_0(\theta,\omega)$ and $b_2(\theta,\omega)$.

In Figs. 3.4 and 3.5, we demonstrate an excellent agreement between theory and simulations for $n(\theta)$ and $p(\theta)$, for given values of (m, T, σ). From the figure, we observe that our analytical approach works very well for both small and large values of m.

The ratio $p(\theta)/n(\theta)$ gives the temperature $T(\theta)$. In the equilibrium state of a system, one necessarily has a spatially uniform temperature profile, i.e., $T(\theta)$ equals the

Fig. 3.4 Density $n(\theta)$ for the dynamics (3.16) in the stationary state for a Gaussian $g(\omega)$, with $m = 0.25$, $T = 0.25$, $\sigma = 0.295$, $k_{\text{trunc}} = 12$ for the upper panel, and $m = 5.0$, $T = 0.25$, $\sigma = 0.2$, $k_{\text{trunc}} = 2$ for the lower panel. The simulations results are denoted by points and are obtained for $N = 10^6$ number of oscillators, while theoretical predictions are denoted by lines. https://doi.org/10.1088/1742-5468/2015/05/P05011 ©*SISSA Medialab Srl. Reproduced by permission of IOP Publishing. All rights reserved.*

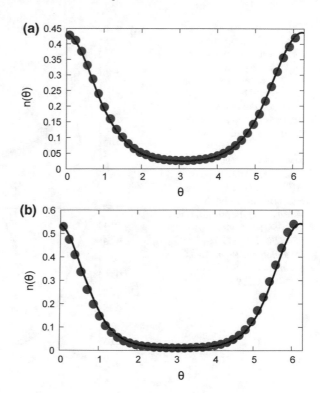

temperature T, independent of θ, where T is the temperature of the heat bath that the system is in equilibrium with. The spatially non-uniform temperature profile shown in the lower panel of Fig. 3.5 is a further evidence of the fact that the synchronized state we are dealing with is a NESS. The figure also shows a density-temperature anticorrelation, i.e., the temperature is peaked at a value of θ at which the density is minimum, and vice versa. This phenomenon of temperature inversion has been argued to be a generic feature of and been observed in long-range interacting systems in NESSs [28–30].

3.4 Phase Diagram: Comparison with Numerics

In this section, we provide a comparison between results for phase transitions obtained from finite-N simulations and theoretical results in the thermodynamic limit discussed above. To this end, let us choose the values of the dimensionless moment of inertia and temperature as $m = 20$ and $T = 0.25$, for which Eq. (3.40) gives $\sigma^{\text{inc}}(m, T) \approx 0.10076$. Our discussion in the preceding section and in particular, our interpretation of the quantity $\sigma^{\text{inc}}(m, T)$ makes us expect the following scenario. For these values of m and T and for a chosen value of σ, on preparing the

Fig. 3.5 In the upper panel is shown the pressure $p(\theta)$ for the same parameters as for the upper panel of Fig. 3.4. Simulation results are depicted by points and correspond to number of oscillators $N = 10^6$, while theoretical predictions are denoted by lines. In the lower panel, we show the local temperature $T(\theta) = p(\theta)/n(\theta)$ and its anticorrelation with the density $n(\theta)$. https://doi.org/ 10.1088/1742-5468/2015/ 05/P05011 ©*SISSA Medialab Srl. Reproduced by permission of IOP Publishing. All rights reserved.*

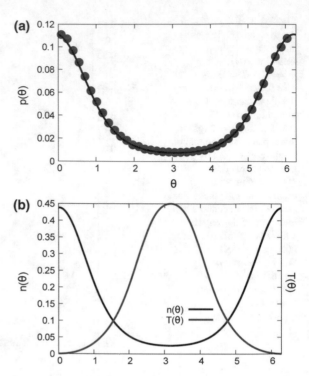

system in an initial state chosen to be the incoherent stationary state at these values of m, T, σ so that r has the initial value of zero, the dynamical evolution will relax r to its stationary-state value corresponding to the synchronized phase, provided that the chosen σ is less than about 0.10076. On the other hand, for σ larger than 0.10076, when the initial incoherent stationary state is linearly stable, r remains zero for all times. We now compare the above continuum-limit predictions with N-body simulations, by monitoring in the latter the evolution of r in time while starting from the incoherent stationary state. To discuss the results from the perspective of statistical physics, let us employ the standard picture of phase transitions as phenomena occurring dynamically due to dissipative relaxation of the order parameter towards the minimum of an underlying phenomenological Landau free energy [31]. For a first-order phase transition, we show schematically in Fig. 3.6 the corresponding schematic free energy $F(r)$ versus r for fixed values of m and T at different σ's. Let us note in passing the very important point that for non-zero σ, one ought to have considered instead of the free-energy the landscapes of the so-called large-deviation functional, which are direct analogs of free-energy in out-of-equilibrium situations [32]. In the discussions pursued here, we assume that the landscape picture of phase

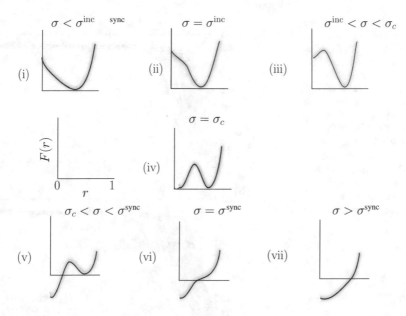

Fig. 3.6 The figure shows schematic Landau free-energy $F(r)$ as a function of r for first-order transitions at fixed m and T while varying σ. A general discussion of forms of such landscapes for first-order transitions may be found in Ref. [23]. Here, panels (**i**) and (**vii**) correspond to the synchronized and the incoherent phase being at the global minimum. In panel (**iii**) (respectively, (**v**)), the synchronized (respectively, incoherent) phase is at the global minimum, while the incoherent (respectively, synchronized) phase is at a local minimum, thus representing a metastable phase. Panel (**iv**) corresponds to the first-order transition point, namely, a point where the synchronized and the incoherent phases coexist at two minima of equal heights

transitions will also hold in the case of large-deviation functions.[3] The utility of invoking the landscape picture lies in its ability to explain, e.g., the flips in r in Fig. 3.3, which correspond to dynamics at σ close to σ_c, when the system switches back and forth between the two almost stable synchronized and incoherent states, see Fig. 3.6iv.

To proceed with the discussion, let us investigate the dynamics for σ around $\sigma^{\text{inc}}(m, T)$. Figure 3.7a–d show simulation results for r as a function of time for four values of σ, two chosen to be below and two above $\sigma^{\text{inc}}(m, T)$ (we have, as mentioned earlier, the values of m and T as $m = 20$ and $T = 0.25$). In each case, we display the dependence of r on time for 10 realizations of the initial incoherent stationary state and for three values of N. Figure 3.7a for $\sigma < \sigma^{\text{inc}}(m, T)$ clearly shows that the system while starting from the unstable incoherent state relaxes in time to the stable synchronized state; this corresponds to a dynamics on the landscape sketched in Fig. 3.6i. The relaxation of r from its initial value to its final synchronized-

---

[3]Obtaining analytical forms of large-deviation functionals for many-body interacting systems turns to be a rather formidable task, and only limited success for very specific model systems has been achieved until now [32].

Fig. 3.7 Panels **a**–**d** show r as a function of time at $m = 20$, $T = 0.25$ for four values of σ, two below (**a**: $\sigma = 0.09$, **b**: $\sigma = 0.095$), and two above (**c**: $\sigma = 0.11$, **d**: $\sigma = 0.12$) the theoretical threshold $\sigma^{\mathrm{inc}}(m, T) \approx 0.10076$, see Eq. (3.40). The data have been obtained by performing N-body simulations of the dynamics (3.16) for a Gaussian $g(\omega)$ with zero mean and unit width

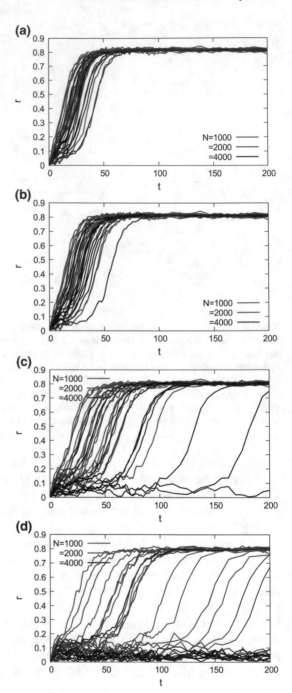

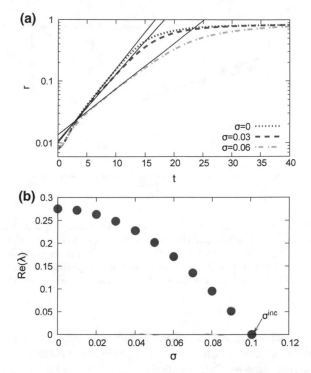

Fig. 3.8 **a** The figure shows simulation results (points) demonstrating exponentially-fast relaxation $\sim e^{\lambda t}$ of r from its initial incoherent state value to its final synchronized state value for $\sigma < \sigma^{\mathrm{inc}}(m, T) \approx 0.10076$ for a Gaussian $g(\omega)$ with $m = 20$, $T = 0.25$, $N = 10^4$. In this figure, the black solid lines denote exponential growth with theoretically computed growth rates λ obtained from Eq. (3.39) for a Gaussian $g(\omega)$ with zero mean and unit width. The simulation data have been obtained by performing N-body simulation of the model (3.16) for a Gaussian $g(\omega)$ with zero mean and unit width. **b** Theoretical λ as a function of σ for the same m and T values as in (**a**); in particular, λ attains the value of zero at the stability threshold $\sigma^{\mathrm{inc}}(m, T)$

state value occurs exponentially fast in time as $e^{\lambda t}$; the growth rate λ is obtained from Eq. (3.39) after substituting a Gaussian distribution for $g(\omega)$. Figure 3.8 demonstrates a match of λ obtained in theory and simulations.

In Fig. 3.7b, plotted for σ larger than its value in (a) but below $\sigma^{\mathrm{inc}}(m, T) \approx 0.10076$, one observes similar to (a) that the system relaxes at long times to the synchronized state for all realizations. This is despite that fact that some of the realizations tend to stay at short times close to the initial incoherent state due to finite-N effects that could not be captured by our continuum-limit theory. For $\sigma > \sigma^{\mathrm{inc}}(m, T)$, but $\sigma < \sigma_c(m, T)$, the landscape sketched in Fig. 3.6iii predicts that the system should relax at long times to the globally stable synchronized state, while should for finite times remain trapped in the metastable incoherent state. This is indeed borne out by the plots in Fig. 3.7c, which shows that while most realizations relax to synchronized states for small N, the number of realizations staying close

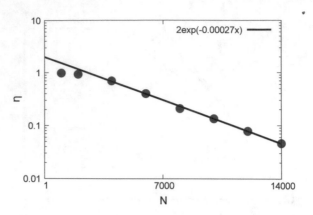

Fig. 3.9 For $m = 20$, $T = 0.25$, $\sigma = 0.11$, the figure shows the fraction η of realizations of initial incoherent state relaxing to synchronized state within the fixed time of observation $t = 200$, for a value of σ above $\sigma^{inc}(m, T)$, for which the incoherent phase is linearly stable in the continuum limit. It is evident from the figure that η for large N decreases exponentially fast with increase of N. The data are obtained in N-body simulations of the dynamics (3.16) for a Gaussian $g(\omega)$ with zero mean and unit width

to the initial incoherent state for a finite time increases with increase of N. It may be shown that the fraction η of realizations relaxing to synchronized state within a fixed and finite time decreases exponentially fast in N for large N, see Fig. 3.9. This plot implies that for the fixed and finite time of observation, a system size larger than those shown in Fig. 3.7c exists for which all realizations stay close to the incoherent state during the dynamical evolution for the time of observation.

A further insight into this last behavior may be obtained by considering the noisy dynamics of a single particle on a potential landscape. In this case, the typical time to get out of a metastable state is given in the weak-noise limit by the Kramers time, i.e., given by a time that is an exponential in the ratio of the potential energy barrier to come out of the metastable state to the strength of the noise [33]. In the case of the dynamics of the order parameter on a free-energy landscape for mean-field systems (similar to the setup of our system), the escape time out of a metastable state obeys Kramers formula with the value of the free-energy barrier replacing the potential energy barrier, but with an additional of N multiplying the barrier height [34]. The factor of N makes the relaxation in mean-field systems quite a slow process compared to short-range systems.[4] It is this last feature that explains the plots in Fig. 3.7c and the behavior of η. Figure 3.7d, plotted for σ larger than $\sigma^{inc}(m, T)$ than in (c), shows that in comparison to (c), more realizations stay close to the initial incoherent state for longer times, since now one has a higher barrier between the incoherent and synchronized state. Based on the above discussions, we may conclude that our theoretical predictions are borne out by our simulation results, but

---

[4]Slow relaxation is a hallmark of long-range interactions, and mean-field interaction is the extreme limit of long-range interaction [15].

one has to exercise caution and correct physical intuition in understanding the finite-N simulation results *vis-à-vis* continuum-limit theory valid in the limit $N \to \infty$. Let us note that the simulation results for $N = 500$ suggest the stability threshold of the incoherent state to lie between $\sigma = 0.095$ and $\sigma = 0.11$, and this range evidently includes its continuum-limit value (≈ 0.10076).

Appendix 1: Proof that the Dynamics (3.16) Does Not Satisfy Detailed Balance

In this appendix, we prove that the dynamics (3.16) does not satisfy detailed balance unless $g(\omega) = \delta(\omega)$, that is, unless σ is zero. For simplicity, we discuss the proof here for the case of two distinct natural frequencies, that is, for a particular bimodal $g(\omega)$ made of two distinct delta peaks. Note that we need at least two different frequencies to have a non-zero σ for the underlying distribution.

Consider a given realization of $g(\omega)$ in which there are N_1 oscillators with frequencies ω_1 and N_2 oscillators with frequencies ω_2, with $N_1 + N_2 = N$. Define the N-oscillator distribution function $f_N(\theta_1, v_1, \ldots, \theta_{N_1}, v_{N_1}, \theta_{N_1+1}, v_{N_1+1}, \ldots, \theta_N, v_N, t)$ as the probability density at time t to observe the system around the values $\{\theta_i, v_i\}_{1 \le i < N}$. In the following, we prefer to use the shorthand notations $z_i \equiv (\theta_i, v_i)$ and $\mathbf{z} = (z_1, z_2, \ldots, z_N)$. Note that f_N satisfies the normalization $\int \left(\prod_{i=1}^N dz_i \right) f_N(\mathbf{z}, t) = 1$. We assume (i) that f_N is symmetric with respect to permutations of dynamical variables within the same group of oscillators, and (ii) that f_N, together with the derivatives $\partial f_N / \partial v_i \, \forall \, i$, vanish on the boundaries of the phase space.

The evolution of f_N follows the Fokker-Planck equation that may be straightforwardly derived from the equations of motion (3.16), to get

$$\frac{\partial f_N}{\partial t} = -\sum_{i=1}^N \left[v_i \frac{\partial f_N}{\partial \theta_i} - \frac{1}{\sqrt{m}} \frac{\partial (v_i f_N)}{\partial v_i} \right] - \sigma \sum_{j=1}^N \left(\Omega^T \right)_j \frac{\partial f_N}{\partial v_j} + \frac{T}{\sqrt{m}} \sum_{i=1}^N \frac{\partial^2 f_N}{\partial v_i^2}$$

$$- \frac{1}{2N} \sum_{i,j=1}^N \sin(\theta_j - \theta_i) \left[\frac{\partial f_N}{\partial v_i} - \frac{\partial f_N}{\partial v_j} \right]. \tag{3.56}$$

Here, the $N \times 1$ column vector Ω is such that its first N_1 entries equal ω_1 and the following N_2 entries equal ω_2, and where the superscript T denotes matrix transpose operation: $\Omega^T \equiv [\omega_1 \, \omega_1 \ldots \omega_1 \, \omega_2 \ldots \omega_2]$.

In order to prove that the dynamics (3.16) does not satisfy detailed balance unless $\sigma = 0$, we first rewrite the Fokker-Planck equation (3.56) as

$$\frac{\partial f_N(\mathbf{x})}{\partial t} = -\sum_{i=1}^{2N} \frac{\partial (A_i(\mathbf{x}) f_N(\mathbf{x}))}{\partial x_i} + \frac{1}{2} \sum_{i,j=1}^{2N} \frac{\partial^2 (B_{i,j}(\mathbf{x}) f_N(\mathbf{x}))}{\partial x_i \partial x_j}, \tag{3.57}$$

where we have

$$x_i = \begin{cases} \theta_i; i = 1, 2, \ldots, N, \\ v_{i-N}; i = N + 1, \ldots, 2N, \end{cases} \tag{3.58}$$

and

$$\mathbf{x} = \{x_i\}_{1 \leq i \leq 2N}. \tag{3.59}$$

In Eq. (3.57), the drift vector $A_i(\mathbf{x})$ is given by

$$A_i(\mathbf{x}) = \begin{cases} v_i; i = 1, 2, \ldots, N, \\ -\frac{1}{\sqrt{m}} v_{i-N} + \frac{1}{N} \sum_{j=1}^{N} \sin(\theta_j - \theta_{i-N}) \\ +\sigma \left(\Omega^T \right)_{i-N}; i = N + 1, \ldots, 2N, \end{cases} \tag{3.60}$$

while the diffusion matrix is

$$B_{i,j}(\mathbf{x}) = \begin{cases} \frac{2T}{\sqrt{m}} \delta_{ij}; i, j > N, \\ 0, \text{ Otherwise.} \end{cases} \tag{3.61}$$

It may be shown that the dynamics described by the Fokker-Planck equation of the form (3.57) satisfies detailed balance if and only if the following conditions are satisfied [35]:

$$\epsilon_i \epsilon_j B_{i,j}(\epsilon \mathbf{x}) = B_{i,j}(\mathbf{x}), \tag{3.62}$$

$$\epsilon_i A_i(\epsilon \mathbf{x}) f_N^s(\mathbf{x}) = -A_i(\mathbf{x}) f_N^s(\mathbf{x}) + \sum_{j=1}^{2N} \frac{\partial B_{i,j}(\mathbf{x}) f_N^s(\mathbf{x})}{\partial x_j}, \tag{3.63}$$

where $f_N^s(\mathbf{x})$ is the stationary solution of Eq. (3.57). Here, $\epsilon_i = \pm 1$ is a constant that denotes the parity with respect to time reversal of the variables x_is: Under time reversal, the latter transform as $x_i \to \epsilon_i x_i$, where $\epsilon_i = -1$ or $+1$ depending on whether x_i is odd or even under time reversal. In our case, θ_is are even, while v_is are odd.

On using Eq. (3.61), we find that the condition (3.62) is trivially satisfied for our model. In order to check the other condition, we formally solve Eq. (3.63) for $f_N^s(\mathbf{x})$, and ask if the solution solves Eq. (3.57) in the stationary state. From Eq. (3.63), we see that for $i = 1, 2, \ldots, N$, the condition is obtained as

$$\epsilon_i A_i(\epsilon \mathbf{x}) f_N^s(\mathbf{x}) = -A_i(\mathbf{x}) f_N^s(\mathbf{x}), \tag{3.64}$$

which on using Eq. (3.60) is obviously satisfied. For $i = N + 1, \ldots, 2N$, we have

$$v_k f_N^s(\mathbf{x}) = -\frac{T \partial f_N^s(\mathbf{x})}{\partial v_k}; \quad k = i - N. \tag{3.65}$$

Solving Eq. (3.65), we get

$$f_N^s(\mathbf{x}) \propto d(\theta_1, \theta_2, \dots, \theta_N) \exp\left[-\tfrac{1}{2T} \sum_{k=1}^N v_k^2\right], \tag{3.66}$$

where $d(\theta_1, \theta_2, \dots, \theta_N)$ is a yet undetermined function. Substituting Eq. (3.66) into Eq. (3.57), and requiring that it is a stationary solution, we arrive at the conclusion that σ has to be equal to zero, and that the factor $d(\theta_1, \theta_2, \dots, \theta_N)$ is given by $d(\theta_1, \theta_2, \dots, \theta_N) = \exp\left(-1/(2NT) \sum_{i,j=1}^N [1 - \cos(\theta_i - \theta_j)]\right)$. Thus, for $\sigma = 0$, when the dynamics reduces to that of the Brownian mean-field model, we get the stationary solution as

$$f_{N,\sigma=0}^s(\mathbf{z}) \propto \exp\left[-\frac{H}{T}\right]. \tag{3.67}$$

where H is the Hamiltonian (expressed in terms of dimensionless variables that were introduced in the main text). The demonstration of the lack of detailed balance for $\sigma \neq 0$ obviously extends to any distribution $g(\omega)$.

Appendix 2: Simulation Details for the Dynamics (3.16)

In this appendix, we describe a method to simulate the dynamics (3.16) for given values of m, T, σ (note that we have dropped the overbars appearing in Eq. (3.16) for not wanting to overload the notation), and for a given realization of ω_i's. We employ a numerical integration scheme discussed in Refs. [5, 6]. Suppose we want to simulate the dynamics over a time interval $[0 : T]$. Let us first choose a time step size $0 < \Delta t \ll 1$. Next, we set $t_n = n\Delta t$ as the n-th time step of the dynamics, where $n = 0, 1, 2, \dots, N_t$, and $N_t = T/\Delta t$. In our numerical scheme, we first discard at every time step the effect of the noise (i.e., consider $1/\sqrt{m} = 0$), and employ a fourth-order symplectic algorithm to integrate the resulting symplectic part of the dynamics [36]. This is followed by adding the effects of noise to the dynamical evolution through implementing an Euler-like first-order algorithm to update the dynamical variables. To summarize, one step of the numerical scheme accounting for evolution between times t_n and $t_{n+1} = t_n + \Delta t$ involves the following updates of the dynamical variables for $i = 1, 2, \dots, N$: For the symplectic part, we have, for $k = 1, \dots, 4$,

$$v_i\left(t_n + \frac{k\Delta t}{4}\right) = v_i\left(t_n + \frac{(k-1)\Delta t}{4}\right) + b(k)\Delta t\left[r\left(t_n + \frac{(k-1)\Delta t}{4}\right)\right.$$
$$\left.\sin\left\{\psi\left(t_n + \frac{(k-1)\Delta t}{4}\right) - \theta_i\left(t_n + \frac{(k-1)\Delta t}{4}\right)\right\} + \sigma\omega_i\right];$$
$$r\left(t_n + \frac{(k-1)\Delta t}{4}\right) = \sqrt{r_x^2 + r_y^2}, \; \psi\left(t_n + \frac{(k-1)\Delta t}{4}\right) = \tan^{-1}\frac{r_y}{r_x},$$

$$r_x = \frac{1}{N} \sum_{j=1}^{N} \sin\left[\theta_j\left(t_n + \frac{(k-1)\Delta t}{4}\right)\right], r_y = \frac{1}{N} \sum_{j=1}^{N} \cos\left[\theta_j\left(t_n + \frac{(k-1)\Delta t}{4}\right)\right],$$

$$(3.68)$$

$$\theta_i\left(t_n + \frac{k\Delta t}{4}\right) = \theta_i\left(t_n + \frac{(k-1)\Delta t}{4}\right) + a(k)\Delta t\, v_i\left(t_n + \frac{k\Delta t}{4}\right), \qquad (3.69)$$

where the constants $a(k)$'s and $b(k)$'s are obtained from Ref. [36] as

$$a(1) = 0.51535283743112293 64, \quad a(2) = 0.085782019412973646,$$
$$a(3) = 0.4415830236164665242, \quad a(4) = 0.1288461583653841854,$$
$$b(1) = 0.13449619927743108 92, \quad b(2) = 0.2248198030794208058,$$
$$b(3) = 0.7563200005156682911, \quad b(4) = 0.3340036032863214255.$$

$$(3.70)$$

At the end of the updates (3.68) and (3.69), we have the set $\{\theta_i(t_{n+1}), v_i(t_{n+1})\}$. We then include the effect of the stochastic noise by keeping the values of the $\theta_i(t_{n+1})$'s unchanged, but by updating $v_i(t_{n+1})$'s as

$$v_i(t_{n+1}) \to v_i(t_{n+1})\left[1 - \frac{1}{\sqrt{m}}\Delta t\right] + \sqrt{2\Delta t \frac{T}{\sqrt{m}}} \Delta X(t_{n+1}). \qquad (3.71)$$

Here, ΔX is a Gaussian distributed random number with zero mean and unit variance.

Appendix 3: Derivation of the Kramers Equation

In this appendix, we derive the Bogoliubov-Born-Green-Kirkwood-Yvon (BBGKY) hierarchy equations for the dynamics (3.16) for any number N of oscillators. This would allow us to then derive in the limit $N \to \infty$ the Kramers equation (3.21) discussed in the main text. Again, as in Appendix 1, we first discuss the derivation of the BBGKY equations for a bimodal $g(\omega)$ made of two distinct delta peaks, and then generalize the derivation to a general $g(\omega)$. Our starting point is the Fokker-Planck equation, Eq. (3.56). To proceed, we follow standard procedure [27], which was also invoked in Chap. 2, and define the so-called reduced distribution function f_{s_1,s_2}, with $s_1 = 0, 1, 2, \ldots, N_1$ and $s_2 = 0, 1, 2, \ldots, N_2$, as

$$f_{s_1,s_2}(z_1, z_2, \ldots, z_{s_1}, z_{N_1+1}, \ldots, z_{N_1+s_2}, t) =$$
$$\frac{N_1!}{(N_1 - s_1)!N_1^{s_1}} \frac{N_2!}{(N_2 - s_2)!N_2^{s_2}} \int dz_{s_1+1} \ldots dz_{N_1} dz_{N_1+s_2+1} \ldots dz_N f_N(z, t). \quad (3.72)$$

Note that the following normalizations hold for the single-oscillator distribution functions: $\int dz_1 f_{1,0}(z_1, t) = 1$, and $\int dz_{N_1+1} f_{0,1}(z_{N_1+1}, t) = 1$.

Using Eq. (3.56) in Eq. (3.72), we get the BBGKY hierarchy equations for oscillators with frequencies ω_1 as

$$\frac{\partial f_{s,0}}{\partial t} + \sum_{i=1}^{s}\left[\frac{v_i \,\partial f_{s,0}}{\partial \theta_i} - \frac{1}{\sqrt{m}}\frac{\partial}{\partial v_i}(v_i f_{s,0})\right] + \sigma \sum_{i=1}^{s}\omega_1 \frac{\partial f_{s,0}}{\partial v_i}$$

$$-\frac{T}{\sqrt{m}}\sum_{i=1}^{s}\frac{\partial^2 f_{s,0}}{\partial v_i^2} = -\frac{1}{2N}\sum_{i,j=1}^{s}\sin(\theta_j - \theta_i)\left[\frac{\partial f_{s,0}}{\partial v_i} - \frac{\partial f_{s,0}}{\partial v_j}\right]$$

$$-\frac{N_1}{N}\sum_{i=1}^{s}\int dz_{s+1}\sin(\theta_{s+1} - \theta_i)\frac{\partial f_{s+1,0}}{\partial v_i}$$

$$-\frac{N_2}{N}\int dz_{N_1+1}\sum_{i=1}^{s}\sin(\theta_{N_1+1} - \theta_i)\frac{\partial f_{s,1}}{\partial v_i}, \tag{3.73}$$

and similar equations for $f_{0,s}$ for oscillators of frequencies ω_2. The first equations of the hierarchy are

$$\frac{\partial f_{1,0}(\theta,v,t)}{\partial t} + \frac{v\partial f_{1,0}(\theta,v,t)}{\partial \theta} - \frac{1}{\sqrt{m}}\frac{\partial}{\partial v}(v f_{1,0}(\theta,v,t))$$

$$+\sigma\omega_1 \frac{\partial f_{1,0}(\theta,v,t)}{\partial v} - \frac{T}{\sqrt{m}}\frac{\partial^2 f_{1,0}(\theta,v,t)}{\partial v^2}$$

$$= -\frac{N_1}{N}\int d\theta' dv' \sin(\theta' - \theta)\frac{\partial f_{2,0}(\theta,v,\theta',v',t)}{\partial v}$$

$$-\frac{N_2}{N}\int d\theta' dv' \sin(\theta' - \theta)\frac{\partial f_{1,1}(\theta,v,\theta',v',t)}{\partial v}, \tag{3.74}$$

and

$$\frac{\partial f_{0,1}(\theta,v,t)}{\partial t} + \frac{v\partial f_{0,1}(\theta,v,t)}{\partial \theta} - \frac{1}{\sqrt{m}}\frac{\partial}{\partial v}(v f_{0,1}(\theta,v,t))$$

$$+\sigma\omega_2 \frac{\partial f_{0,1}(\theta,v,t)}{\partial v} - \frac{T}{\sqrt{m}}\frac{\partial^2 f_{0,1}(\theta,v,t)}{\partial v^2}$$

$$= -\frac{N_2}{N}\int d\theta' dv' \sin(\theta' - \theta)\frac{\partial f_{0,2}(\theta,v,\theta',v',t)}{\partial v}$$

$$-\frac{N_1}{N}\int d\theta' dv' \sin(\theta' - \theta)\frac{\partial f_{1,1}(\theta,v,\theta',v',t)}{\partial v}. \tag{3.75}$$

In the limit of large N, we may write

$$g(\omega) = \left[\frac{N_1}{N}\delta(\omega - \omega_1) + \frac{N_2}{N}\delta(\omega - \omega_2)\right], \tag{3.76}$$

and express Eqs. (3.74) and (3.75) in terms of $g(\omega)$.

To generalize Eqs. (3.74) and (3.75) to the case of a continuous $g(\omega)$, we denote for this case the single-oscillator distribution function as $f(\theta, v, \omega, t)$. The first equation of the hierarchy is then obtained as

$$
\frac{\partial f(\theta, v, \omega, t)}{\partial t} + \frac{v \partial f(\theta, v, \omega, t)}{\partial \theta} - \frac{1}{\sqrt{m}} \frac{\partial}{\partial v}(v f(\theta, v, \omega, t))
$$

$$
+ \sigma \omega \frac{\partial f(\theta, v, \omega, t)}{\partial v} - \frac{T}{\sqrt{m}} \frac{\partial^2 f(\theta, v, \omega, t)}{\partial v^2}
$$

$$
= - \int d\omega' g(\omega') \int d\theta' dv' \sin(\theta' - \theta) \frac{\partial f(\theta, v, \theta', v', \omega, \omega', t)}{\partial v}. \qquad (3.77)
$$

In the continuum limit $N \to \infty$, we may neglect two-oscillator correlations (as we have done in Chap. 2) and approximate $f(\theta, v, \theta', v', \omega, \omega', t)$ as

$$
f(\theta, v, \theta', v', \omega, \omega', t) = f(\theta, v, \omega, t) f(\theta', v', \omega', t), \qquad (3.78)
$$

neglecting terms that are sub-dominant in N. Using the last equation in Eq. (3.77) reduces the latter to the Kramers equation (3.21), thereby achieving the goal of this appendix.

Appendix 4: Nature of Solutions of Eq. (3.39)

In this appendix, we analyze in detail the nature of solutions of Eq. (3.39). We rewrite the equation as

$$
F(\lambda; m, T, \sigma) \equiv \frac{e^{mT}}{2T} \sum_{p=0}^{\infty} \frac{(-mT)^p \left(1 + \frac{p}{mT}\right)}{p!}
$$

$$
\times \int \frac{g(\omega) d\omega}{1 + \frac{p}{mT} + \frac{\lambda}{T\sqrt{m}} + i\frac{\sigma\omega}{T}} - 1 = 0, \qquad (3.79)
$$

where $g(\omega)$ is unimodal. The incoherent state will be unstable if there is a λ with a positive real part that satisfies the above eigenvalue equation. We will now prove that depending on the parameters appearing in the above equation, there can be at most one such λ that can be only real. Moreover, for the case of a Gaussian $g(\omega)$, we will obtain the general shape of the surface in the (m, T, σ) space that defines the instability region of the incoherent state.

Considering m and T strictly positive, we multiply for convenience the numerator and denominator of Eq. (3.79) by mT to arrive at

$$F(\lambda; m, T, \sigma) = \frac{e^{mT}}{2T} \sum_{p=0}^{\infty} \frac{(-mT)^p (p+mT)}{p!}$$

$$\times \frac{g(\omega)d\omega}{mT + p + \sqrt{m}\lambda + i\sigma m\omega} - 1 = 0. \tag{3.80}$$

We now look for pure imaginary solutions of this equation. Separating into real and imaginary parts the last equation, we have

$$\mathrm{Re}\,[F(i\mu; m, T, \sigma)] = \frac{e^{mT}}{2T} \sum_{p=0}^{\infty} \frac{(-mT)^p}{p!}$$

$$\times \int d\omega\, g(\omega) \frac{(p+mT)^2}{(p+mT)^2 + \left(m\sigma\omega + \sqrt{m}\mu\right)^2} - 1 = 0, \tag{3.81}$$

$$\mathrm{Im}\,[F(i\mu; m, T, \sigma)] = -\frac{e^{mT}}{2T} \sum_{p=0}^{\infty} \frac{(-mT)^p}{p!}$$

$$\times \int d\omega\, g(\omega) \frac{(p+mT)\left(m\sigma\omega + \sqrt{m}\mu\right)}{(p+mT)^2 + \left(m\sigma\omega + \sqrt{m}\mu\right)^2} = 0. \tag{3.82}$$

In the second equation above, let us make the change of variables $m\sigma\omega + \sqrt{m}\mu = m\sigma x$, and exploit the parity in x of the sum. We get

$$\mathrm{Im}\,[F(i\mu; m, T, \sigma)] =$$

$$-m\sigma \int_0^\infty dx \left\{ \left[g\left(x - \frac{\mu}{\sqrt{m\sigma}}\right) - g\left(-x - \frac{\mu}{\sqrt{m\sigma}}\right) \right] \right.$$

$$\left. \times x \sum_{p=0}^{\infty} \frac{(-mT)^p}{p!} \frac{p+mT}{(p+mT)^2 + m^2\sigma^2 x^2} \right\} = 0, \tag{3.83}$$

where it may be shown that the sum on the right-hand side is positive definite for any finite σ. Furthermore, for the class of $g(\omega)$ considered in the main text, one may see that the term in square brackets is positive (respectively, negative) definite for $\mu > 0$ (respectively, for $\mu < 0$). As a result, the last equation is never satisfied for $\mu \neq 0$ and finite, and therefore, the eigenvalue equation does not admit pure imaginary solutions (the proof holds also for the particular case $g(\omega) = \delta(\omega)$, as may be checked).

We may also conclude that there can be at most one solution with positive real part. In fact, if in the complex λ-plane, we consider the loop depicted in Fig. 3.10, panel (a) (where the points A and C represent $\mathrm{Im}\lambda \to \pm\infty$, respectively, and the radius of the arc extends to ∞), we obtain correspondingly in the complex-$F(\lambda)$ plane due to the sign properties of $\mathrm{Im}\,[F(i\mu; m, T, \sigma)]$ just described the loop represented schematically in Fig. 3.10, panel (b). The point $F = -1$ in panel (b) is obtained for λ in panel (a) at points A and C and in the whole of the arc extending to infinity. The position of the point B in the complex-F plane is determined by the value of $F(0)$,

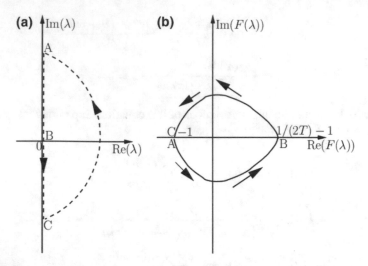

Fig. 3.10 The loop in the complex F-plane, (**b**), corresponding to the loop in the complex λ-plane, (**a**), as determined by the function $F(\lambda)$ given in Eq. (3.80)

which is given by

$$F(0; m, T, \sigma) = \frac{e^{mT}}{2T} \sum_{p=0}^{\infty} \frac{(-mT)^p}{p!}$$

$$\times \int d\omega \, g(\omega) \frac{(p + mT)^2}{(p + mT)^2 + (m\sigma\omega)^2} - 1. \qquad (3.84)$$

From the well-known theorem of complex analysis on the number of roots of a function in a given domain of the complex plane [37], we arrive at the result that for $F(0; m, T, \sigma) > 0$, there is one and only one solution of the eigenvalue equation with positive real part; on the other hand, for $F(0; m, T, \sigma) < 0$, there is no such solution. When the single solution with positive real part exists, it is necessarily real, since a complex solution would imply the existence of its complex conjugate. The value of $F(0; m, T, \sigma)$ may be seen to equal $1/(2T) - 1$ for $\sigma = 0$. For positive σ, the value will depend on the specific form of the distribution function $g(\omega)$. However, one can prove that the value is always smaller than $1/(2T) - 1$, consistent with the physically reasonable fact that if the incoherent state is stable for $\sigma = 0$, which happens for $T > \frac{1}{2}$, it is all the more stable for $\sigma > 0$.

The surface delimiting the region of instability in the (m, T, σ) phase space is implicitly defined by Eq. (3.84) (i.e. $F(0; m, T, \sigma) = 0$), which in principle can be solved to obtain the threshold value of σ (denoted by σ^{inc}) as a function of (m, T): $\sigma^{\text{inc}} = \sigma^{\text{inc}}(m, T)$. On physical grounds, we expect that the latter is a single valued function, and that for any given value of m, it is a decreasing function of T for $0 \le T \le 1/2$, reaching 0 for $T = 1/2$. We are able to prove analytically these facts for the class of unimodal distribution functions $g(\omega)$ considered in the main text that includes the Gaussian case. However, we can prove for any $g(\omega)$ that $\sigma^{\text{inc}}(m, T)$ tends to 0 for $m \to \infty$. This is done using the integral representation

$$\sum_{p=0}^{\infty} \frac{(-mT)^p}{p!} \frac{(p+mT)^2}{(p+a)^2 + (m\sigma\omega)^2} = e^{-mT}$$

$$- (m\sigma\omega) \int_0^{\infty} dt \, \exp\left[-mT\left(t + e^{-t}\right)\right] \sin(m\sigma\omega t); \tag{3.85}$$

For $\sigma > 0$ and $m \to \infty$, one may see that the term within the integral in the last equation tends to e^{-mT}. We thus obtain by examining Eq. (3.84) that $F(0; m \to \infty, T > 0, \sigma > 0) = -1$. Combined with the fact that $F(0; m, T, 0) = 1/(2T) - 1$, this shows that $\sigma^{\text{inc}}(m \to \infty, 0 \le T \le \frac{1}{2}) = 0$.

We now turn to the Gaussian case, $g(\omega) = 1/\sqrt{2\pi} \exp\left[-\omega^2/2\right]$. Denoting with a subscript g this case, and using Eq. (3.85), we have

$$F_g(0; m, T, \sigma) = \frac{1}{2T} - 1 - \frac{e^{mT}}{2T\sqrt{2\pi}} \int d\omega \, e^{-\frac{\omega^2}{2}} (m\sigma\omega)$$

$$\int_0^{\infty} dt \, \exp\left[-mT\left(t + e^{-t}\right)\right] \sin(m\sigma\omega t). \tag{3.86}$$

The integral in ω may be easily performed: Making the change of variable $m\sigma t = y$, we arrive at the equation

$$F_g(0; m, T, \sigma) = \frac{1}{2T} - 1$$

$$- \frac{1}{2T} \int_0^{\infty} dy \, y e^{-\frac{y^2}{2}} \exp\left[mT\left(1 - \frac{y}{m\sigma} - e^{-\frac{y}{m\sigma}}\right)\right]. \tag{3.87}$$

The equation $F_g(0; m, T, \sigma) = 0$ defines implicitly the function $\sigma^{\text{inc}}(m, T)$, which we can show to be a single-valued function with the properties $\frac{\partial \sigma^{\text{inc}}}{\partial m} < 0$ and $\frac{\partial \sigma^{\text{inc}}}{\partial T} < 0$. We show these by explicitly computing the partial derivatives of $F_g(0; m, T, \sigma)$ with respect to m and σ, and by evaluating the behavior with respect to changes in T by adopting a suitable strategy.

We begin by computing the derivative with respect to σ. From Eq. (3.87), we obtain

$$\frac{\partial}{\partial \sigma} F_g(0; m, T, \sigma) = -\frac{1}{2\sigma^2} \int_0^{\infty} dy \, y^2 e^{-\frac{y^2}{2}} \left(1 - e^{-\frac{y}{m\sigma}}\right)$$

$$\times \exp\left[mT\left(1 - \frac{y}{m\sigma} - e^{-\frac{y}{m\sigma}}\right)\right], \tag{3.88}$$

which is clearly negative. Secondly, the derivative with respect to m gives

$$\frac{\partial}{\partial m} F_g(0; m, T, \sigma) = -\frac{1}{2} \int_0^\infty dy \, y e^{-\frac{y^2}{2}}$$
$$\times \left(1 - e^{-\frac{y}{m\sigma}} - \frac{y}{m\sigma} e^{-\frac{y}{m\sigma}} \right) \exp\left[mT \left(1 - \frac{y}{m\sigma} - e^{-\frac{y}{m\sigma}} \right) \right]. \tag{3.89}$$

This derivative is negative, since $1 - e^{-x} - xe^{-x}$ is positive for $x > 0$. From the implicit function theorems, we then derive the result $\frac{\partial \sigma^{\mathrm{inc}}}{\partial m} < 0$. The study of the behavior with respect to a change in T is more complicated. Since we are considering $T > 0$, we multiply Eq. (3.87) by $2T$ to obtain

$$2T F_g(0; m, T, \sigma) = 1 - 2T$$
$$- \int_0^\infty dy \, y e^{-\frac{y^2}{2}} \exp\left[mT \left(1 - \frac{y}{m\sigma} - e^{-\frac{y}{m\sigma}} \right) \right]. \tag{3.90}$$

Consider the integral on the right-hand side

$$\int_0^\infty dy \, y e^{-\frac{y^2}{2}} \exp\left[mT \left(1 - \frac{y}{m\sigma} - e^{-\frac{y}{m\sigma}} \right) \right]; \tag{3.91}$$

Since $1 - x - e^{-x}$ is negative for $x > 0$, we conclude that the T derivative of this expression is negative, while its second T derivative is positive. Then the right-hand side of Eq. (3.90) can be zero for $T > 0$ for at most one value of T. Furthermore, since for fixed y and m, the value of $y/(m\sigma)$ decreases if σ increases, we conclude that the T value for which $F_g(0; m, T, \sigma) = 0$ decreases for increasing σ at fixed m. This concludes the proof. Furthermore, for what we have seen before, we have $\sigma^{\mathrm{inc}}(m, 1/2) = 0$ and $\lim_{m \to \infty} \sigma^{\mathrm{inc}}(m, T) = 0$ for $0 \le T \le 1/2$.

From the above analysis, it should be evident that the proof is not restricted to the Gaussian case, but would work for any $g(\omega)$ such that

$$\beta \int dx \, g(x) x \sin(\beta x), \tag{3.92}$$

is positive for any β. However, on physical grounds, we are led to assume that the same conclusions hold for any unimodal $g(\omega)$.

Appendix 5: Solution of the System of Eqs. (3.46)

In this appendix, we give details of the solution, Eqs. (3.47)–(3.53), to the system of Eqs. (3.46).

Let us first consider Eq. (3.46) with $n = 0$, from which we obtain that $c_{1,k}(\theta, \omega)$ is independent of θ for each k. Next, consider Eq. (3.46) for $k = 0$ and $n = 2, 3, \ldots$. Since we have $c_{n,0}(\theta, \omega) = 0$ for $n > 0$, we find that $c_{n,1}(\theta, \omega) = 0$ for $n > 1$. Next, Eq. (3.46) for $k = 1$ and $n = 3, 4, \ldots$ gives $c_{n,2}(\theta, \omega) = 0$ for $n > 2$; for $k = 2$ and

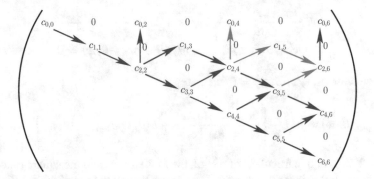

Fig. 3.11 Flow diagram for the evaluation of the expansion coefficients $c_{n,k}(\theta, \omega)$; $n, k = 0, 1, 2, \ldots, 6$ by using Eq. (3.46). Starting from the main diagonal, arrows and different colors denote subsequent flows (see text). The elements below the main diagonal are all zero. https://doi.org/10.1088/1742-5468/2015/05/P05011 ⓒSISSA Medialab Srl. Reproduced by permission of IOP Publishing. All rights reserved.

$n = 4, 5, \ldots$, we get $c_{n,3}(\theta, \omega) = 0$ for $n > 3$, and so on. We are thus led to conclude that $c_{n,k}(\theta, \omega) = 0 \, \forall \, k < n$. Figure 3.11 displays the coefficients $c_{n,k}$ in a matrix that is seen to be upper triangular on the basis of the result just obtained. Hence, to obtain all the non-zero elements of the matrix, we should consider Eq. (3.46) for $n = 1, 2, \ldots$ and $k \geq n - 1$, or, equivalently, for $k = 0, 1, 2, \ldots$ and $n = 1, 2, \ldots, k + 1$. To this end, we will first obtain the elements of the main diagonal, $c_{n,n}(\theta, \omega)$, then the elements of the first upper diagonal, $c_{n,n+1}(\theta, \omega)$, then the elements of the second upper diagonal, $c_{n,n+2}(\theta, \omega)$, and so on.

Let us begin by studying the case of $n = 1$ and $k = 0$; we have

$$\sqrt{T}\frac{\partial c_{0,0}(\theta, \omega)}{\partial \theta} + \sqrt{2T}\frac{\partial c_{2,0}(\theta, \omega)}{\partial \theta} + \sqrt{T}a(\theta, \omega)c_{0,0}(\theta, \omega) + c_{1,1}(\omega) = 0. \quad (3.93)$$

Here, we have $c_{2,0}(\theta, \omega) = 0$, while $c_{1,1}(\omega)$ is independent of θ. We thus end up with a first-order differential equation for $c_{0,0}(\theta, \omega)$ with an unknown constant. The condition $c_{0,0}(\theta, \omega) = c_{0,0}(\theta + 2\pi, \omega)$ fixes the value of this constant, and we get

$$c_{0,0}(\theta, \omega) = c_{0,0}(0, \omega)e^{-g(\theta,\omega)}\left[1 + \left(e^{g(2\pi,\omega)} - 1\right)\frac{\int_0^\theta d\theta' e^{g(\theta',\omega)}}{\int_0^{2\pi} d\theta' e^{g(\theta',\omega)}}\right], \quad (3.94)$$

$$c_{1,1}(\omega) = \sqrt{T}\frac{c_{0,0}(0, \omega)\left(1 - e^{g(2\pi,\omega)}\right)}{\int_0^{2\pi} d\theta' e^{g(\theta',\omega)}}, \quad (3.95)$$

where $g(\theta, \omega) = \int_0^\theta d\theta' a(\theta', \omega)$, and $c_{0,0}(0, \omega)$ has to be fixed at the end by the normalization of $b_0(\theta, \omega)$. Now that we have determined $c_{0,0}(\theta, \omega)$ and $c_{1,1}(\omega)$, we may obtain recursively the main diagonal elements by considering Eq. (3.46) for $n = 2, 3, \ldots$ and $k = n - 1$; we get

$$\sqrt{nT}\frac{\partial c_{n-1,n-1}(\theta,\omega)}{\partial\theta} + \sqrt{(n+1)T}\frac{\partial c_{n+1,n-1}(\theta,\omega)}{\partial\theta}$$

$$+\sqrt{nT}a(\theta,\omega)c_{n-1,n-1}(\theta,\omega) + nc_{n,n}(\theta,\omega) = 0. \tag{3.96}$$

Since we have $c_{n+1,n-1}(\theta,\omega) = 0$, we get

$$c_{n,n}(\theta,\omega) = -\sqrt{\frac{T}{n}}\left[\frac{\partial c_{n-1,n-1}(\theta,\omega)}{\partial\theta} + a(\theta,\omega)c_{n-1,n-1}(\theta,\omega)\right] \tag{3.97}$$

for $n = 2, 3, \ldots$. In particular, for $n = 2$, the first term within the square brackets is absent as $c_{1,1}(\omega)$ is independent of θ. Let us note that all the functions $c_{n,n}(\theta,\omega)$ are proportional to $c_{0,0}(0,\omega)$.

Now, we determine the elements of the first upper diagonal. Consider Eq. (3.46) for $n = 1$ and $k = 1$:

$$\sqrt{T}\frac{\partial c_{0,1}(\theta,\omega)}{\partial\theta} + \sqrt{2T}\frac{\partial c_{2,1}(\theta,\omega)}{\partial\theta} + \sqrt{T}a(\theta,\omega)c_{0,1}(\theta,\omega) + c_{1,2}(\omega) = 0. \tag{3.98}$$

The last equation has exactly the same structure as Eq. (3.93), since $c_{2,1}(\theta,\omega) = 0$, and $c_{1,2}(\omega)$ is a constant independent of θ. Now, we use the fact that $c_{0,k}(0,\omega) = 0$ for $k \geq 1$, so that the solution of Eq. (3.98) is simply $c_{0,1}(\theta,\omega) = c_{1,2}(\omega) \equiv 0$. Next, by considering Eq. (3.46) for $n = 2, 3, \ldots$ and $k = n$, and proceeding similarly, we may obtain that all the functions $c_{n,n+1}(\theta,\omega)$, i.e., the elements of the first upper diagonal of Fig. 3.11, are zero.

Our next task is to determine the elements of the second upper diagonal, which we begin by considering Eq. (3.46) for $n = 1$ and $k = 2$:

$$\sqrt{T}\frac{\partial c_{0,2}(\theta,\omega)}{\partial\theta} + \sqrt{2T}\frac{\partial c_{2,2}(\theta,\omega)}{\partial\theta} + \sqrt{T}a(\theta,\omega)c_{0,2}(\theta,\omega) + c_{1,3}(\omega) = 0. \tag{3.99}$$

In the above equation, $c_{2,2}(\theta,\omega)$ is known from Eq. (3.97). Then, from the requirement of periodicity of $c_{0,2}(\theta,\omega)$, and on using $c_{0,2}(0,\omega) = 0$, we arrive at the solutions

$$c_{0,2}(\theta,\omega) = \sqrt{2}\frac{\int_0^{2\pi} d\theta' \frac{\partial c_{2,2}(\theta',\omega)}{\partial\theta'} e^{g(\theta',\omega)}}{\int_0^{2\pi} d\theta' e^{g(\theta',\omega)}} e^{-g(\theta,\omega)} \int_0^{\theta} d\theta' e^{g(\theta',\omega)}$$

$$-\sqrt{2}e^{-g(\theta,\omega)} \int_0^{\theta} d\theta' \frac{\partial c_{2,2}(\theta',\omega)}{\partial\theta'} e^{g(\theta',\omega)}, \tag{3.100}$$

$$c_{1,3}(\omega) = -\sqrt{2T}\frac{\int_0^{2\pi} d\theta' \frac{\partial c_{2,2}(\theta',\omega)}{\partial\theta'} e^{g(\theta',\omega)}}{\int_0^{2\pi} d\theta' e^{g(\theta',\omega)}}. \tag{3.101}$$

Again, these functions are proportional to $c_{0,0}(0,\omega)$. Now that we have determined $c_{0,2}$ and $c_{1,3}$, we obtain recursively the elements of the second upper diagonal, i.e., the functions $c_{n,n+2}$, from Eq. (3.46) by considering $n = 2, 3, \ldots$ and $k = n + 2$:

$$\sqrt{nT}\frac{\partial c_{n-1,n+1}(\theta,\omega)}{\partial\theta} + \sqrt{(n+1)T}\frac{\partial c_{n+1,n+1}(\theta,\omega)}{\partial\theta}$$

$$+\sqrt{nT}a(\theta,\omega)c_{n-1,n+1}(\theta,\omega) + nc_{n,n+2}(\theta,\omega) = 0. \tag{3.102}$$

With the main diagonal elements already determined, we get

$$c_{n,n+2}(\theta,\omega) = -\sqrt{\frac{T}{n}}\left[\frac{\partial c_{n-1,n+1}(\theta,\omega)}{\partial\theta} + a(\theta,\omega)c_{n-1,n+1}(\theta,\omega)\right]$$

$$-\frac{\sqrt{(n+1)T}}{n}\frac{\partial c_{n+1,n+1}(\theta,\omega)}{\partial\theta}, \tag{3.103}$$

for $n = 2, 3, \ldots$. In particular, for $n = 2$, the first term within the square brackets is absent, as $c_{1,3}(\omega)$ is independent of θ. Also, note that these functions are proportional to $c_{0,0}(0,\omega)$.

We now show that the elements of the third upper diagonal vanish. Considering Eq. (3.46) for $n = 1$ and $k = 3$, we get

$$\sqrt{T}\frac{\partial c_{0,3}(\theta,\omega)}{\partial\theta} + \sqrt{2T}\frac{\partial c_{2,3}(\theta,\omega)}{\partial\theta} + \sqrt{T}a(\theta,\omega)c_{0,3}(\theta,\omega) + c_{1,4}(\omega) = 0. \tag{3.104}$$

Here, $c_{2,3}$ has been previously determined to be vanishing identically, so that the solution of the last equation is simply obtained as $c_{0,3}(\theta,\omega) = c_{1,4}(\omega) \equiv 0$. Then, considering Eq. (3.46) for $n = 2, 3, \ldots$ and $k = n+2$, we conclude that all the elements of the third upper diagonal, $c_{n,n+3}$, vanish.

By now, the procedure of determining the coefficients $c_{n,k}$'s should be clear. All the elements of the upper diagonals of odd order vanish, being equivalent to the fact that in the portion of each row above the main diagonal, one element every two vanishes, i.e., $c_{n,n+1+2k} \equiv 0$ for $n, k = 0, 1, 2, \ldots$. All the nonvanishing elements are proportional to $c_{0,0}(0,\omega)$. The expressions for the main diagonal elements are given by Eqs. (3.94), (3.95) and (3.97). On the basis of the analysis above, we may write down the general expressions for the nonvanishing non-diagonal elements as

$$c_{0,2k}(\theta,\omega) = \sqrt{2}\frac{\int_0^{2\pi} d\theta' \frac{\partial c_{2,2k}(\theta',\omega)}{\partial\theta'} e^{g(\theta',\omega)}}{\int_0^{2\pi} d\theta' e^{g(\theta',\omega)}}e^{-g(\theta,\omega)}\int_0^\theta d\theta' e^{g(\theta',\omega)}$$

$$-\sqrt{2}e^{-g(\theta,\omega)}\int_0^\theta d\theta'\frac{\partial c_{2,2k}(\theta',\omega)}{\partial\theta'}e^{g(\theta',\omega)}, \tag{3.105}$$

$$c_{1,1+2k}(\omega) = -\sqrt{2T}\frac{\int_0^{2\pi} d\theta' \frac{\partial c_{2,2k}(\theta',\omega)}{\partial\theta'} e^{g(\theta',\omega)}}{\int_0^{2\pi} d\theta' e^{g(\theta',\omega)}}, \tag{3.106}$$

$$c_{2,2+2k}(\theta,\omega) = -\sqrt{\frac{T}{2}}a(\theta,\omega)c_{1,1+2k}(\omega) - \frac{\sqrt{3T}}{2}\frac{\partial c_{3,1+2k}(\theta,\omega)}{\partial\theta}, \tag{3.107}$$

$$c_{n,n+2k}(\theta,\omega) = -\sqrt{\frac{T}{n}}\left[\frac{\partial c_{n-1,n-1+2k}(\theta)}{\partial\theta} + a(\theta,\omega)c_{n-1,n-1+2k}(\theta,\omega)\right]$$

$$-\frac{\sqrt{(n+1)T}}{n}\frac{\partial c_{n+1,n-1+2k}(\theta,\omega)}{\partial\theta} \qquad n\geq 3, \qquad (3.108)$$

with $k = 1, 2, \ldots$.

Appendix 6: Convergence Properties of the Expansion (3.45)

In this appendix, we discuss the convergence properties of the expansion (3.45) involved in obtaining the density $n(\theta)$, see Eq. (3.54). To this end, let us first consider an asymptotic power series in the real variable x given by

$$A(x) = \sum_{k=0}^{\infty} a_k x^k. \qquad (3.109)$$

We define the partial sum

$$A_n(x) \equiv \sum_{k=0}^{n} a_k x^k. \qquad (3.110)$$

Then, being asymptotic means that at any given $x \neq 0$, one has $A_n(x) \to \infty$ as $n \to \infty$. In this case, one employs the so-called Borel summation method to sum the series, by defining the Borel transform of $A(x)$ as [26]

$$\mathscr{B}A(t) \equiv \sum_{k=0}^{\infty} \frac{a_k}{k!} t^k. \qquad (3.111)$$

If $\mathscr{B}A(t)$ converges for any positive t, or, if it converges for sufficiently small t to an analytic function that can be analytically continued to all $t > 0$, and if the integral

$$\int_0^{\infty} dt \, \exp(-t)\mathscr{B}A(tx) \qquad (3.112)$$

exists and equals $A_B(x)$ (here, the subscript B stands for Borel), we say that the Borel sum of the series on the right hand side of Eq. (3.109) is $A_B(x)$. One may observe that if the original series converges, i.e., if $\lim_{n\to\infty} A_n(x) = A(x) < \infty$, then one has $A_B(x) = A(x)$. Applying the above formalism to Eq. (3.45), we get

$$b_{0B}(\theta,\omega) = \int_0^{\infty} dt \, \exp(-t) \sum_{k=0}^{\infty} \frac{c_{0,k}(\theta,\omega)}{k!} (t\sqrt{m})^k$$

$$= \frac{1}{\sqrt{m}} \int_0^{\infty} dy \, \exp(-y/\sqrt{m}) \sum_{k=0}^{\infty} \frac{c_{0,k}(\theta)}{k!} y^k. \qquad (3.113)$$

Fig. 3.12 Density $n(\theta)$ in
the dynamics (3.16) for a
Gaussian $g(\omega)$, and with
$m = 0.25$, $T = 0.25$,
$\sigma = 0.295$. Panel **a** refers to
theoretical predictions using
the Borel summation method
with $k_{\text{trunc}} = 38$, while **b**
refers to estimates obtained
by using direct summation
with $k_{\text{trunc}} = 22$

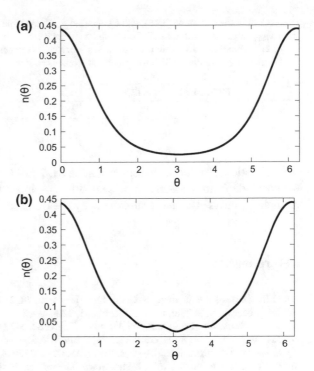

The last integral needs to be computed numerically. One is required to truncate the
series at a certain order $k = k_{\text{trunc}}$, and to extend the integral over y up to a given
value y_M, chosen such that the integrand is negligible for $y > y_M$. However, contrary
to what happens in the original series, we found that the sum in the last integral
converges, at least for all y-values smaller than y_M that are necessary to compute the
integral. We do not know the function to which our Borel transform converges and
the corresponding radius of convergence. Nevertheless, our numerical results show
that our series is Borel summable. Figure 3.12a shows the result of computing the
density

$$n(\theta) = \int_{-\infty}^{\infty} d\omega\, g(\omega) b_{0B}(\theta, \omega) \qquad (3.114)$$

for the same conditions as in Fig. 3.4, where we truncate the sum in Eq. (3.113) at
$k_{\text{trunc}} = 38$; the plot coincides with the one shown in Fig. 3.4a. On the other hand, summing the series (3.45) for $n = 0$ without resorting to the Borel summation method,
and then computing the density $n(\theta)$, the result that we show in Fig. 3.12b clearly
demonstrates that instabilities for truncation order $k_{\text{trunc}} = 22$ of the series that get
worse and worse with further increase of the truncation order. In this regard, the
reader is referred to Table 3.1, where we list the truncation order k_{max} versus m
up to which one observes a perfect agreement of the density $n(\theta)$ between theory

Table 3.1 For the dynamics (3.16) with a Gaussian $g(\omega)$, the table shows the maximum truncation order k_{max} in the computation of the density $n(\theta)$ as a function of m at a given representative $(\sigma, T) \equiv (0.295, 0.25)$ for which one observes a perfect agreement of the density $n(\theta)$ in theory and simulations, as in Fig. (3.4). The agreement worsens on increasing the truncation order beyond k_{max}

m	0.0625	0.125	0.25	0.5	1.0
k_{max}	60	32	18	10	6

and simulations, for the same representative $(\sigma, T) \equiv (0.295, 0.25)$ as in Fig. (3.4). We conclude from the analysis presented in this appendix that the series (3.45) is asymptotic, but is effectively summable by the Borel summation method.

References

1. H. Tanaka, A.J. Lichtenberg, S. Oishi, Phys. Rev. Lett. **78**, 2104 (1997)
2. J.A. Acebrón, R. Spigler, Phys. Rev. Lett. **81**, 2229 (1998)
3. J.A. Acebrón, L.L. Bonilla, R. Spigler, Phys. Rev. E **62**, 3437 (2000)
4. J.A. Acebrón, L.L. Bonilla, C.J.P. Vicente, F. Ritort, R. Spigler, Rev. Mod. Phys. **77**, 137 (2005)
5. S. Gupta, A. Campa, S. Ruffo, Phys. Rev. E **89**, 022123 (2014)
6. S. Gupta, A. Campa, S. Ruffo, J. Stat. Mech. Theory Exp. R08001 (2014)
7. M. Komarov, S. Gupta, A. Pikovsky, EPL **106**, 40003 (2014)
8. S. Olmi, A. Navas, S. Boccaletti, A. Torcini, Phys. Rev. E **90**, 042905 (2014)
9. A. Campa, S. Gupta, S. Ruffo, J. Stat. Mech. Theory Exp. P05011 (2015)
10. J. Barré, D. Métivier, Phys. Rev. Lett. **117**, 214102 (2016)
11. P.H. Chavanis, Eur. Phys. J. B **87**, 120 (2014)
12. M. Antoni, S. Ruffo, Phys. Rev. E **52**, 2361 (1995)
13. A. Campa, T. Dauxois, S. Ruffo, Phys. Rep. **480**, 57 (2009)
14. F. Bouchet, S. Gupta, D. Mukamel, Physica A **389**, 4389 (2010)
15. A. Campa, T. Dauxois, D. Fanelli, S. Ruffo, *Physics of Long-Range Interacting Systems* (Oxford University Press, Oxford, 2014)
16. Y. Levin, R. Pakter, F.B. Rizzato, T.N. Teles, F.P.C. Benetti, Phys. Rep. **535**, 1 (2014)
17. S. Gupta, S. Ruffo, Int. J. Mod. Phys. A **32**, 1741018 (2017)
18. B. Ermentrout, J. Math. Biol. **29**, 571 (1991)
19. H. Sakaguchi, Prog. Theor. Phys. **79**, 39 (1988)
20. R. Livi, P. Politi, *Nonequilibrium Statistical Physics: A Modern Perspective* (Cambridge University Press, Cambridge, 2017)
21. G. Filatrella, A.H. Nielsen, N.F. Pedersen, Eur. Phys. J. B **61**, 485 (2008)
22. M. Rohden, A. Sorge, M. Timme, D. Witthaut, Phys. Rev. Lett. **109**, 064101 (2012)
23. N. Goldenfeld, *Lectures on Phase Transitions and the Renormalization Group* (Addison-Wesley, Reading, 1992)
24. H. Risken, *The Fokker-Planck Equation: Methods of Solution and Applications* (Springer, Berlin, 1996)
25. N.N. Lebedev, *Special Functions and their Applications* (Dover, New York, 1972)
26. G.H. Hardy, *Divergent Series* (Chelsea, New York, 1991)
27. K. Huang, *Statistical Mechanics* (Wiley, New York, 1987)
28. L. Casetti, S. Gupta, Eur. Phys. J. B **87**, 91 (2014)
29. T.N. Teles, S. Gupta, P. Di Cintio, L. Casetti, Phys. Rev. E **92**, 020101(R) (2015)

30. S. Gupta, L. Casetti, New J. Phys. **18**, 103051 (2016)
31. K. Binder, Rep. Prog. Phys. **50**, 783 (1987)
32. H. Touchette, Phys. Rep. **478**, 1 (2009)
33. H.A. Kramers, Physica VII **4**, 284 (1940)
34. R.B. Griffiths, C.Y. Weng, J.S. Langer, Phys. Rev. **149**, 301 (1966)
35. C.W. Gardiner, *Handbook of Stochastic Methods for Physics, Chemistry and the Natural Sciences* (Springer, Berlin, 1983)
36. R.I. McLachlan, P. Atela, Nonlinearity **5**, 541 (1992)
37. V.I. Smirnov, *A Course of Higher Mathematics. Vol. 3. Part. 2, Complex Variables Special Functions* (Pergamon Press, Oxford, 1964) Chap. 1, Sec. 22

Printed in the United States
By Bookmasters